# Mountains and Lowlands

*Ancient Iran and Mesopotamia*

# Mountains and Lowlands

## *Ancient Iran and Mesopotamia*

PAUL COLLINS

ASHMOLEAN

ASHMOLEAN



British Library Cataloguing-in-Publication Data
*A catalogue record for this book is available from the British Library*

ISBN 978 1 910807 08 8



Published by the Ashmolean Museum, University of Oxford
General Editor: Katherine Wodehouse
Designed and typeset in Gill Sans Nova and Fournier by Adrian Hunt
Printed and bound in Belgium by Albe de Coker

For further details of Ashmolean titles please visit:
www.ashmolean.org/shop

THIS PAGE:
Detail of fig.23, siltstone 'Two-dog' palette, from Hierakonpolis, Egypt, *c.*3200–3100 BC

UNIVERSITY OF
OXFORD

# Contents

# FOREWORD

AT A TIME WHEN the heritage of large areas of the Middle East is under threat as never before, it is especially important that the achievements of this region's great civilisations are understood and celebrated. Today, the Ashmolean is one of the few places in the world where it is possible to trace the stories of those achievements though objects that span over 10,000 years: from the earliest farming communities to the world's first cities and the great empires that unified the lands reaching from Egypt and the Aegean to the Indus Valley and Central Asia. The Museum's Ancient Near East collections therefore encompass not only a vast span of time but also an enormous and varied geographical region containing a mosaic of diverse peoples, languages, and cultures. This book explores the archaeology and history of two areas, Iran and Mesopotamia, tracing their interconnections through remarkable objects and works of art across six millennia.

The Ashmolean Museum has played a leading role in the field of ancient Near Eastern archaeology for over a century, supporting excavations at some 52 sites across Iran, Iraq, Israel, Jordan, Palestine, and Turkey. One of the most important collections in the Museum is arguably that derived

from the archaeological work conducted jointly by the University of Oxford and the Field Museum, Chicago, at the site of Kish in southern Iraq between 1923 and 1933. As was the practice at the time, finds were divided between the host country and the funding institutions. The Ashmolean received nearly 3,000 objects ranging in date from the late fourth millennium BC to the seventh century AD. Among the discoveries were hundreds of clay documents written in the wedge-shaped cuneiform script of ancient Mesopotamia: these, together with examples of tablets from other sites, means that the Museum now holds the second largest collection of cuneiform texts in the United Kingdom, including examples of some of the world's earliest writing. Once in Oxford, the finds were available for study and it was principally through decades of scholarly work by Dr Roger Moorey (1937–2004) of the Ashmolean's Department of Antiquities that the material from Kish became known to both scholars and the wider public. Among his other achievements were the cataloguing of the Museum's important collections of Iranian (Luristan) metalwork and ancient Near Eastern terracottas and stamp seals. Such work is continued by the many researchers from around the globe who annually visit the Museum in order to study and publish the collections, helping to reconstruct the societies that lie at the root of our own urban, literate world.

In the production of the book thanks are due to David Gowers for his superb photographic work and to Katherine Wodehouse and Declan McCarthy for guiding it through the editorial and design process.

We are also profoundly grateful to Jaleh Hearn who has both supported the curatorial work on the ancient Near East collections and generously funded this publication.

XA STURGIS
Director, Ashmolean Museum

ABOVE:
Detail of fig.32, modern impression of a cylinder seal with hero and wild animals, from Kish, Iraq, *c.*2700–2600 BC

OPPOSITE:
Detail of fig.67, bronze disc-headed pin with grotesque face and animals, *c.*950–650 BC

# 1. Setting the Scene

'Skidding goat' pottery vessel,
*c.*3200 BC, fig.21 (detail)

# 1. Setting the Scene

THIS BOOK IS INTENDED to provide students and the general reader with an introduction to the archaeological and political histories of Iran and Mesopotamia (modern Iraq and eastern Syria) in the period 6000 BC–AD 650, that is, from a time when the entire region had become the home to scattered farming, fishing, and herding communities to its conquest by Arab armies and the introduction of Islam. Many previous studies as well as museum exhibitions have tended to explore Mesopotamia's achievements with an eye on the Mediterranean world, often as part of a search for the foundations of Western civilisation. This book, however, looks eastwards from the Mesopotamian lowlands and westwards from the plateau of Iran, using as its axis the great Zagros Mountain chain – stretching from the Taurus range in the north to the head of the Persian Gulf and then further to the south-east – that both connects and separates the two areas (fig.1). The aim is to shift the focus away from an emphasis on Mesopotamia as the source of all cultural developments and to acknowledge the active influence of Iran – and indeed to suggest that the extraordinary accomplishments of the ancient Near East would not have been possible without the long history of tension and collaboration between the inhabitants of the uplands and of the lowlands. The essential differences between Iran and Mesopotamia are that the former was rich in a wide variety of resources such as stones, metals, and wood but the rugged nature of the landscape meant that only a limited number of areas were suitable for intensive settlement and these were irregularly distributed. In contrast, Mesopotamia was rich in limited, but fundamental, resources with fertile soils, wide plains, and easy communication (fig.2). The story that unfolds here is the result of the complex interplay between the populations of these two regions and their utilisation of their different resources.

The Zagros was a permeable membrane rather than a barrier, but, even before modern political borders were imposed (the present border between Iraq and Iran has its origins in a treaty of 1639 between the Ottoman and Safavid Empires), invisible boundaries existed between the populations of the mountains and those of the plains in the form of cultural differences – languages, religions, forms of production, and lifestyles as well as social and political organisations. However, the routes through the mountains that brought materials such as lapis lazuli and copper, and animals such as horses, from Iran to Mesopotamia were also networks that facilitated the spread of meanings, values, ideas, technologies, and people in both directions.

**FIG. 1** A fertile valley within the southern Zagros Mountains © Wolfgang Kaehler/Corbis

This interaction contributed to the evolution of communities from their earliest stages and there is no doubt that the hilly flanks of the Zagros were a heartland in the fundamental transition from mobile lifestyles of hunting and foraging to sedentary farming and animal herding, together with – as modern excavations are revealing – associated cultic and complex funerary behaviour. This book enters the story at the start of the sixth millennium BC when cultural connections were being established over ever wider regions.

## ROUTES OF CONTACT AND CONFLICT

Between September 1980 and August 1988 one of the longest and bloodiest armed conflicts of the twentieth century was fought between the countries of Iran and Iraq. At the cost of perhaps a million lives, the eight-year war

**FIG. 2** The alluvial plain of Mesopotamia with date plantations near Babylon © Namir Noor-Eldeen/Pool/epa/Corbis

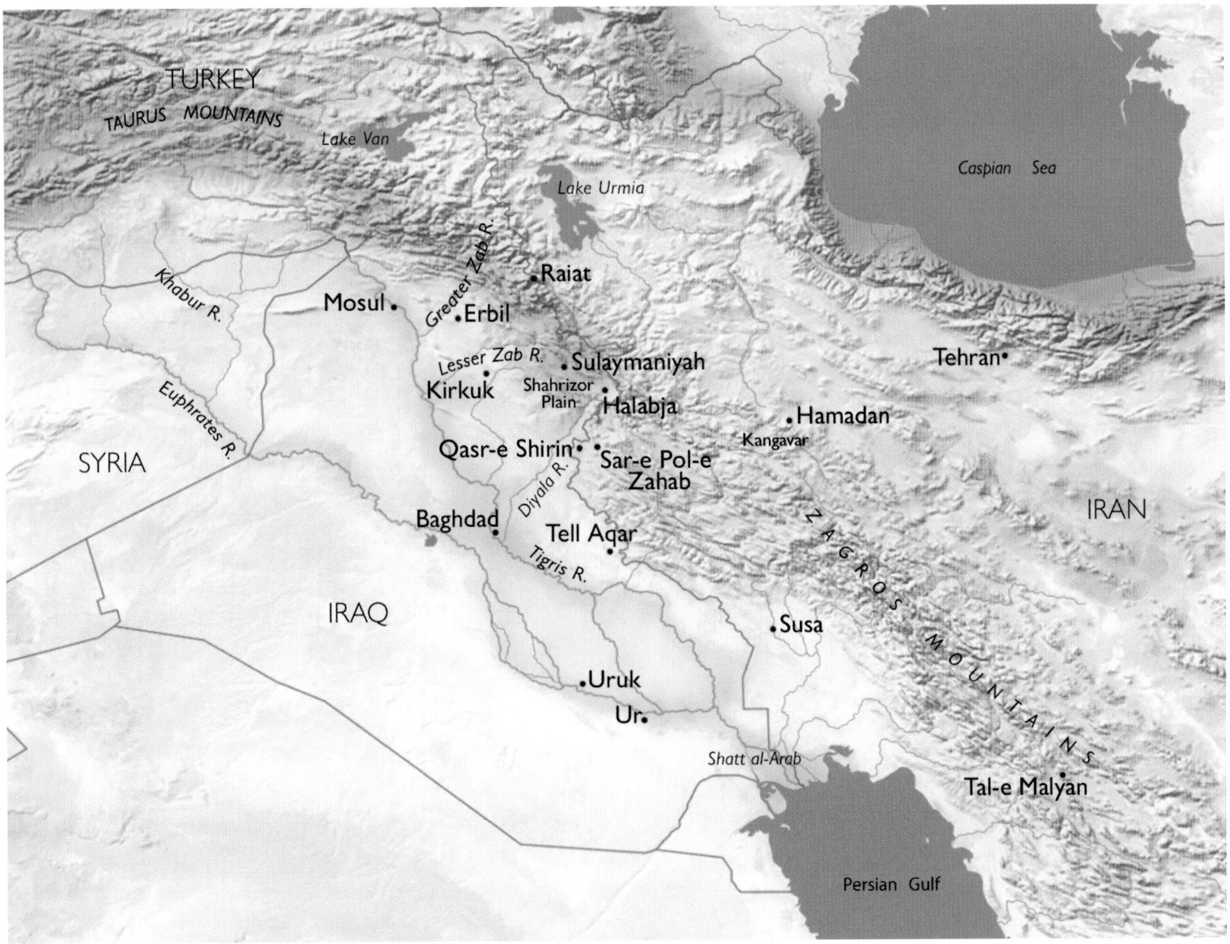


between the revolutionary Shi'ite regime of Ayatollah Khomeini and the Sunni dictatorship of Saddam Hussein was fought back and forth along the 1,460 kilometre border. Fighting was concentrated in three main areas: Kurdistan in the north; the central region east of the Diyala River; and in the far south along the Shatt al-Arab (or Arvand Rud) waterway and the adjoining Iranian province of lowland Khuzestan. These regions had already played an equally strategic role in the ancient relations between Iran and Mesopotamia as within them lay the principal routes through the Zagros used by armies, merchants, pilgrims, and pastoralists (map 1, above).

In the north, the mountains of Kurdistan can be passed at two crucial points: from Mosul (ancient Nineveh) and Erbil (ancient Arbela) a route follows the Greater Zab River via Raiat towards Lake Urmia; and from Kirkuk (ancient Arrapha) through the Sulaymaniyah Plain to Halabja and south-eastwards to join the highway that is described next.

From central Mesopotamia the River Diyala provides a passageway to the region of Qasr-e Shirin, from where a route turns south-east via Sar-e Pol-e Zahab and then east through the Mahidasht and Kangavar valleys to the city of Hamadan (ancient Ecbatana), and the Iranian Plateau beyond.

This remains today an important road through the Zagros and represents the western end of the so-called Silk Route or Great Khorasan Road (the name adopted for the roadway here).

A fourth road, further south, ran eastwards from the Tigris and then parallel with the Zagros from Tell Aqar (ancient Der) along the foothills via the Deh Luran Plain to Susa in Khuzestan. Routes from the Susiana plains then led through the mountains, one of the most significant being the plain–foothill–highland corridor connecting Susa with Tal-e Malyan (ancient Anshan) to the south-east.

Of equal importance in this story is the so-called piedmont route running along the slopes of the foothills of the Zagros east of the Tigris that connected with tracks in the north leading across the plains of upper Mesopotamia and into the mountains of Anatolia (modern Turkey and Armenia) beyond. This became the Great Royal Road, one of the most important arteries of the Persian Empire during the fifth to fourth centuries BC, but even by that date it had been a long-established highway linking the western and eastern regions of Western Asia, from Sardis (in western Anatolia) to Susa. The precise route taken and the crossing-points of rivers would probably have varied depending on the seasons and the prevailing political situation, but a traveller from Syria would have crossed the steppe either north or south of the Jebel Sinjar and reached the Tigris at a river-crossing above Nineveh. The route could then have passed either side of the Jebel Maqlub and so across the Greater Zab River to Arbela before continuing south-east alongside the Zagros foothills across the Lesser Zab to Arrapha. Reaching the Diyala, roads would have allowed the traveller to head westwards onto the alluvial plain of southern Mesopotamia or east along the Great Khorasan Road to Ecbatana.

In any discussion of routes connecting Mesopotamia and Iran it is important also to highlight the Persian Gulf. The lowland plains of Susiana in southern Khuzestan lie adjacent and are broadly similar to the alluvial plains of southern Mesopotamia, but the two are geographically distinct. Indeed they appear to have been separated by the headwaters of the Persian Gulf throughout the fifth and fourth millennia BC. Thus contact between the two would have involved a long and circuitous overland route north from places such as Ur and Uruk, and then south-east via the piedmont: the most direct and cheapest route would have been by boat across the Persian Gulf.

## RECONSTRUCTING THE PAST

Imagine an old-fashioned film reel that has disintegrated over time into a number of torn and damaged strips of various lengths. These have been scattered around a room and many sections have been lost. The challenge is to reconstruct the reel and make sense of the resulting film. Through careful study of the form and content of each strip of film, it becomes possible to determine sequences that can be spliced together. The film is silent: at the start there are no subtitles but occasional written words appear later – though only for a few of the many actors and sometimes in languages we barely understand – which help to make the action a little more meaningful. With hundreds of missing frames, however, it is necessary to be creative in order to construct a narrative, perhaps by adding footage from other films. The opinion of the critics may well be divided over the result.

Such is the challenge for archaeologists and historians working with the fragmentary remains of the past in Mesopotamia and Iran. Archaeological work has been patchy and conducted over the last century and a half with varying degrees of skill and publication. The distribution of sites reflects the distribution of archaeological work – vast regions have not been explored, or certainly not with modern archaeological methods. Some areas are well known and thus appear, perhaps unfairly, to play a more significant role in the narrative than might really be justified. Although, as will be seen, complex forms of recording were developed in southern Mesopotamia and south-western Iran in the fourth millennium BC, it was not until around 2500 BC that written texts begin to allow us to say something of the history of the region. Even then the reconstruction of events is often indebted to royal inscriptions and these cannot be considered as reliable 'historical sources'. This means that attempting to write a narrative history of the early Near East is virtually impossible. Nonetheless, at a general level it is possible to provide a survey which highlights the interrelationships between regions, peoples, and political states. These connections were often driven by economic concerns and the demand for resources; the movement and acquisition of these resources therefore plays a central role in the account. Such links also influenced the cultural development of various societies as revealed by their material remains, especially the objects of the elite who benefited most obviously from these socio-political exchanges.

# II. From Village to City: 6000–3000 BC

Limestone cylinder seal with silver ram handle, *c.*3200–3000 BC, fig.19

# II. From Village to City: 6000–3000 BC

BY 6000 BC, SMALL VILLAGE communities were scattered throughout much of Mesopotamia and within the foothills and fertile valleys of the Zagros Mountains (map 2, right). People had settled, sometimes for many generations, in areas of good farmland and in places where resources were abundant or easily accessible along natural land routes and watercourses. This included the vast delta wetlands at the head of the Persian Gulf where life was based on the herding of cattle, although sheep and goats were also important, together with the cultivation of wheat, barley, and linseed using simple gravity-flow irrigation from the multiple branches of the rivers Tigris and Euphrates. The rivers and lagoons were a source of fish while the extensive marshes supplied reeds in unlimited quantities for a range of purposes including the creation of mats, baskets, rafts, and even houses; these were all made watertight using naturally occurring bitumen that was obtained from seepages in Susiana to the south-east (fig.3). One of the increasingly important resources of the region may already have been the date palm – the earliest evidence for its exploitation and possible cultivation comes from sites of the late sixth or early fifth millennium BC. The fruits of this tree, harvested in late summer, were probably dried and stored and, together with dried fish, created a completely adequate diet throughout

FIG. 3 Two women poling a *mashuf* canoe in the marshes of southern Iraq, photograph taken in 1951, 2004.130.29532.1 © Pitt Rivers Museum, University of Oxford

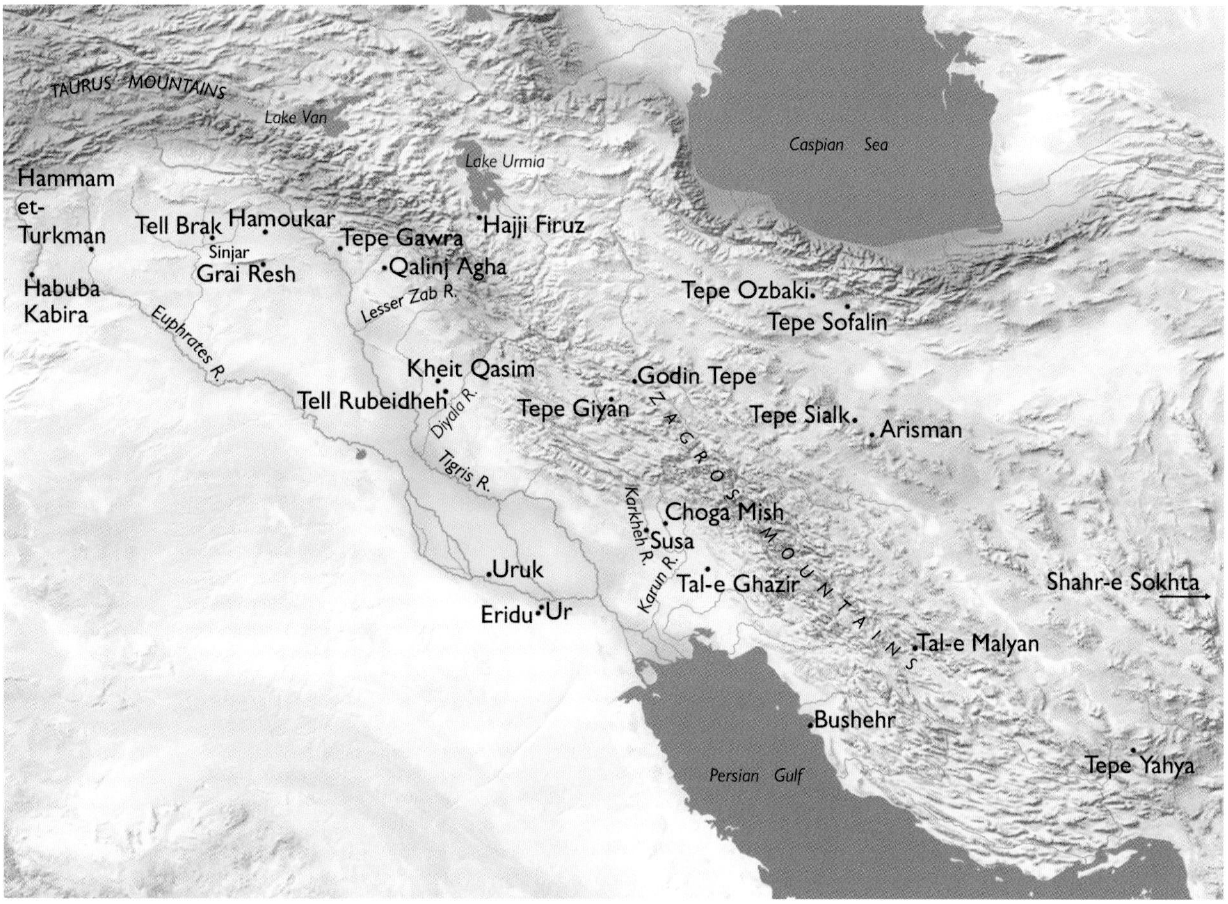


the year. In addition, the palm sprout could have been eaten and the sap used for sweetening and fermenting into an alcoholic drink. Nearly every part of the tree had a practical use: date stones were an effective fuel; the trunk was sufficient to make light buildings, boats, and furniture, while the fibre could be twisted to make ropes; temporary shelters could be made from the fronds and individual leaves woven into baskets and mats.

The herding of cattle was also an important activity across the northern plains of Mesopotamia along with the rain-fed cultivation of wheat and barley (though irrigation was used to increase yields). In regions east of the Tigris and within the valleys of the Zagros, sheep and goats predominated together with the farming of some cereals. The highlands were covered by oak woodland and the intervening steppe with pistachio and almond trees as well as the wild ancestors of apricots, plums, pomegranates, and grapes. Here, as elsewhere across much of Mesopotamia and Iran, mud or mudbrick houses offered families spaces for sleeping, eating, and working, as well as protection from wild animals and the extremes of weather (fig.4). However, life was precarious. It always remained possible that there would be insufficient rain for cultivation, that the cereals would fail to ripen, that the reserves of food stored in baskets and clay containers would rot or be eaten by rodents, or that some sheep and cattle, or even worse, family

FIG. 4 Mudbrick houses in the mountains of the northern Zagros, photograph taken in 1951, 2004.130.32142.1 © Pitt Rivers Museum, University of Oxford

FIG. 5 Halaf-style painted plate, from Chagar Bazar, Syria, *c.*5500–5000 BC, diam. 22 cm, AN1936.79 and Ubaid-style spouted vessel, from Ur, Iraq, *c.*4500–4000 BC, ht. 15.5 cm, AN1935.783 © Ashmolean Museum, University of Oxford

members, would die as a result of accident, disease, or animal attack. In such a situation, individual and group rituals and social activities would have played a significant role in maintaining a sense of meaning, identity, and cohesion. From at least the middle of the sixth millennium BC, a feature in these regular events might have been the serving and drinking of wine. Evidence for the world's oldest known grape wine was discovered in six jars, each with a volume of about nine litres, set into the floor of a small house dating to around 5400 BC at the site of Hajji Firuz in the north-western Zagros. Remarkably, the colours of the preserved residues suggest that there were both red and white varieties from which to choose. The domestication of wild grapes and the beginning of winemaking almost certainly took place within the mountainous zones reaching from the Taurus to the Zagros, from where the technology and consumption would have spread to the lowlands. The drinking of such beverages would play a central role in ritual and feasting within societies across ancient Iran and Mesopotamia throughout the period covered by this book. Techniques of fermentation were almost certainly not limited to wine. By this period wheat and barley had already been domesticated and cultivated across the region for millennia and the fermentation of cereal with wild yeasts to create beer may have already been long recognised. Iran can claim the earliest chemically confirmed barley beer at the site of Godin Tepe dating to around 3200 BC, although there is possible evidence for its production at Gobekli Tepe in south-eastern Anatolia as early as the ninth millennium BC. These fermented drinks were much preferred to water: they are generally safer than untreated water, more nutritious and, of course, can have psychotropic effects.

Such drinks might have been served in the handmade pottery vessels that were a characteristic feature of village life throughout the lowlands and highlands at this time. Different areas across the region produced ceramics with their own distinct shapes and elaborately painted designs; these were modified over time to reflect changing interests and influences. Across northern Mesopotamia, roughly within the dry farming region from the Zagros to the steppe lands west of the Euphrates, thin-walled and finely decorated pottery of the so-called Halaf tradition was produced and exchanged (fig.5). Within the Zagros there was a diverse range of regional styles. In contrast, a distinctive Ubaid-style of black-on-buff pottery was made in the villages of the southern alluvium and by the end of the sixth millennium BC there were strong parallels between the pottery produced in southern Mesopotamia and that of the communities of the Susiana plains and along the Persian Gulf coast as far as the area of Bushehr.

Around 5000 BC the highly decorated vessels of northern Mesopotamia were giving way to the simpler styles that had developed in the south. However, there was a pronounced degree of variation in the ways that ceramics were used and understood across this wider region, which reflected the area's many local lifestyles. Pastoralists, for example, moving with their sheep, goats, or cattle, would have used pottery in a markedly different way to those in settled farming societies. The spread of a common pottery style can possibly be explained by the wide adoption of the slow wheel technology used to produce it – perhaps encouraged by itinerant potters – rather than by movements of large numbers of people from southern Mesopotamia into areas to the north and south-east.

Evolving social practices within villages across the wider region may also have promoted this spread, as well as allowing other aspects of the Ubaid traditions to be adopted in the north. This included one of the most distinctive features of settled communities in Mesopotamia: the layout of ordinary mudbrick houses that took the form of a rectangular room flanked on each long side by a number of smaller rooms (fig.6). This tripartite arrangement may have provided appropriate accommodation for extended families. The central room, perhaps unroofed, was probably used for eating and cooking, and was the place for receiving guests as well as for ritual activities. The side rooms were for storage, sleeping, and the production

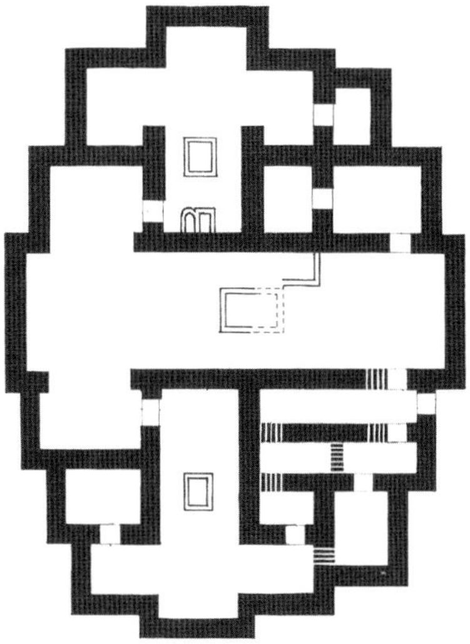

**FIG. 6** Plan of an Ubaid period house at Kheit Qasim, Iraq. After C.K. Maisels, *The Near East: Archaeology in the 'Cradle of Civilization'* (London & New York, 1993)

of pottery, tools, and the processing of foods. Such organised spaces may have allowed for household activities to become more focused resulting in greater productivity within the community. These new ways in which people organised their lives were also probably being encouraged by their acquisition and processing of novel materials such as metals and wool. The spinning and weaving of wool to create blankets and clothing created a new class of objects for both household use and exchange and, with patterns woven through them, textiles could become important markers of identity (perhaps replacing some of the designs found earlier on pottery). Space in the home for this work had to be defined, while out in the pasturelands, the management of herds now involved new skills such as the shearing of sheep as much as their milking and slaughter. With these developments came new roles within the community.

Materials and ideas moved in a complex network that tied settled populations to mobile groups, enabled by the seasonal movements of animals from lowland to highland pastures and vice versa. This led to other forms of shared ideas. In the Iranian highlands, as revealed by excavated human skulls, there had been a long tradition of head shaping for a small number of individuals. The shaping of the malleable skulls of babies, using textiles, wooden boards, or even simply by massaging the head, can be undertaken without any known side effects. The result is a flattened or elongated shaped head. This practice was adopted across the lowlands during the sixth and fifth millennia BC. Significantly, some clay figurines of humans from sites in both Iran and Mesopotamia of this date are shown with elongated or exaggerated head shapes (see fig.8).

## AN INCREASE IN SCALE

Shared approaches to the acquisition, production, and exchange of materials resulted in the emergence of settlements that served as focal points for wider communities. A few such sites dating from around 4500–4200 BC have been identified by archaeologists. These are often situated on the boundaries between areas with differing resources, lying at the intersections of lowlands, mountains, and waterways. Some settlements played an important role in rituals, others were small market towns where manufacturing took place. Members of the community (perhaps individual families or extended groups) may have accrued sufficient authority to mobilise workers at these sites to create significant architecture in the form of temples – the status of these people was thus validated through a relationship with the gods.

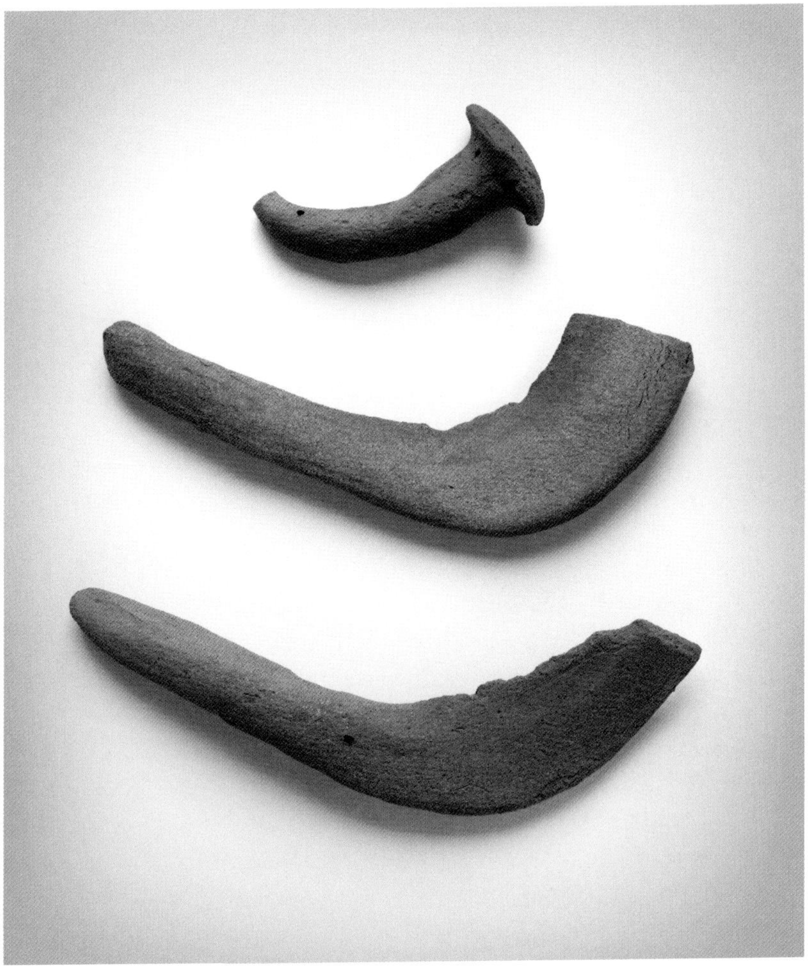

**FIG. 7** Baked clay sickles, l. 22.5 cm, AN1925.656 and baked clay 'nail', l. 10 cm, AN1925.657a, from Eridu and Tell al-Ubaid, Iraq, *c.*5500–4000 BC © Ashmolean Museum, University of Oxford

One such place was Eridu, positioned on, or at least very near to, the coast, at the head of the Persian Gulf (which was much further north than it is today). This was a place where maritime activity met the agricultural and herding world of southern Mesopotamia. By this date the technology of boat building was such that sea-going vessels could travel considerable distances along the coastline for fishing and the acquisition of exotic resources. Ubaid-style pottery found on the shores of the Persian Gulf, to the north and south of the Qatar Peninsula, was perhaps carried there by people from places like Eridu who established camps or visited the local inhabitants. Inland from Eridu lay a world of reed marshes, natural river channels, and human-made canals that criss-crossed the rich alluvial soil of southern Mesopotamia. Reed and palm boats and rafts carried people, animals, and products easily and cheaply between settlements (see fig.3). The reeds, as well as other canal-bank plants and cereals, were cut using clay sickles, a tool highly characteristic of southern Mesopotamia and a cheap alternative to flint sickles (fig.7). Made in vast numbers using moulds, they were fired in kilns along with pottery. The sickles are often found with nail-shaped objects also made of baked clay. The pointed end is often bent over and the head end shows signs of wear; the 'nails' were perhaps used in food preparation or in the processing of skins or reeds.

**FIG. 8** Baked clay figurines of women, from Ur, Iraq, *c.*5500–4000 BC, ht. 13.8 cm, 1930,1213.173
© Trustees of the British Museum

The settlement at Eridu was focused on a large mudbrick platform measuring about eighteen by eleven metres that supported a temple building. In layout the temple mirrored the tripartite arrangement of an ordinary house but was constructed on a much larger scale. The platform and temple went through a series of re-buildings. With an old mudbrick structure it is often cheaper and easier to demolish rather than repair it, packing the lower courses with the rubble to form the foundations for a new building – over several centuries this results in raising a building ever higher above the succession of earlier structures. At Eridu there were perhaps six re-buildings of the platform and temple so that the later versions would have been above the roofs of ordinary houses, dominating the skyline. The facade of the temple itself was decorated with recesses and buttresses, the presence of which was to become a standard way of defining monumental mudbrick architecture in Mesopotamia for millennia to come. The central hall was entered through a door in the eastern long wall. Sufficient evidence survived in the hall of the later temples for archaeologists to identify a free-standing pedestal and another clay structure, perhaps an altar, set against the southern short wall. Some of the rituals associated with the home were now likely to have been undertaken in the temple on behalf of the wider community. Over time access to the god may have become more exclusive, and those with authority could claim rights of access either directly or through a specialised group of priests. Rituals and religious beliefs probably acted to underpin community relations both locally and more widely. That the Eridu temple was a place of ritual offerings may be indicated by the quantities of fish bones found in a chamber adjoining the central hall. Indeed, written records of the third millennium BC inform us that Eridu was the cult centre for Enki, god of sweet water whose symbol was a fish, and it is not impossible that he was also the focus for prehistoric worship at the site.

Some 50 metres to the south-west of the temple area lay a large communal cemetery with approximately 1,000 graves of which around 200 have been excavated. The burials were in rectangular pits, lined with mudbricks. The deceased were all laid on their backs, with arms to the side or with hands folded across the pelvis. Similar burials of the same date are also known from Ur, approximately twenty kilometres to the north-east. Some of the graves at both sites contained baked clay human figurines (fig.8). These have distinctive 'coffee-bean' shaped eyes, and elongated heads with headdress or hair indicated with bitumen, and rounded pellets of clay applied to the shoulders.

In contrast to the Mesopotamian plains where cattle seem to have played an important role, on the lowlands of Susiana there was a focus on the herding of sheep and goats. In the middle of the fifth millennium BC, the largest site in the central plain was Chogha Mish (covering an area of approximately fifteen hectares). It was surrounded by smaller settlements that were centres for the production of stone tools, pottery – similar in style to that used in Mesopotamia – and the extraction of bitumen. These activities were perhaps being controlled from Chogha Mish where a monumental building of this period has been excavated. The building was eventually destroyed by fire after which the settlement was abandoned. Some of the population of Chogha Mish, along with that of surrounding villages, may have moved west and settled at the site of Susa. Whether this was by force or by choice is unknown. Nonetheless, the result was that Susa soon emerged as an important centre for the wider region. Situated between the Karkheh and Karun rivers, the town was surrounded by fertile, easily irrigated soils with access to the Persian Gulf as well as to the resources of the Zagros. From approximately 4300 to 4150 BC, Susa sat at the apex of a network of smaller sites, acting as a political, economic, or religious centre around which different and shifting communities of villages, herding camps, and craft producers gravitated.

There were two main areas of settlement at Susa, named by their excavators as the Acropole and the Apadana mounds, the latter encircled by a six-metre-wide mud wall. It was the Acropole that was the most significant area of the town. Perhaps as early as 4200 BC, a low platform of unbaked mudbrick was built here. Although its size was not impressive, covering an area only about seven by fourteen metres, this was the focus for a vast cemetery with perhaps as many as 2,000 individual burials, apparently interred over a relatively short period. It is possible that these were the inhabitants of Susa and neighbouring sites whose remains had been gathered together from a number of smaller cemeteries or charnel houses. It is also possible, however, that they are the consequence of catastrophic disease, famine, or conflict. As the site was first excavated in 1906–8 and poorly recorded, a conclusive explanation may never be possible. What makes this find all the more extraordinary is that the deceased were interred with extremely fine and elaborately painted handmade ceramics (fig.9). The unique character of these objects was the result of influences from the types of contemporary ceramics used within the Zagros. Scientific analysis of the vessels has shown that they were produced at several sites close to Susa. It is possible that as many as 8,000 vessels were deposited in the cemetery, perhaps manufactured not for daily use but as prestige goods for the grave. They fall into three clear types: beakers, dishes, and jars, which may reflect three types of food and drink thought necessary for the deceased. Other exotic items found in the burials included copper axes and discs, highlighting the increasing importance of this metal within society.

Alongside the small mudbrick platform, a much larger solid stepped terrace, nearly eleven metres high, was constructed. A two-metre-high base supported a nine-metre-high platform covering an area of approximately 82 by 90 metres (slightly smaller than a modern football pitch, it was certainly considerably larger than the contemporary temple platform at Eridu). The sides of the platform were decorated with inlaid ceramic cones, arranged into groups of four or five, as well as clay plaques and models of goat horns. On top of the platform were structures that may have been for grain storage and workshops but might also have included temple buildings.

It was not only in the south of the region that settlements had emerged as the focus for smaller and scattered communities. In the hilly country of north-eastern Mesopotamia lay Tepe Gawra. Along with Eridu and Susa, it shared many of the Ubaid traditions in the second half of the fifth millennium BC. Gawra lies between the Tigris River and the first foothills of the Zagros Mountains. The site is close to the entrance of an important pass that leads onto the Iranian Plateau through the Jebel Maqlub. That this

served as a transport link for goods and raw materials carried by human porters – since no pack animals had yet been domesticated – from the highlands and Upper Tigris basin into Mesopotamia is indicated by the presence of pieces of lapis lazuli at the site. At this time the only source for this exquisite blue stone was Badakhshan, the mountainous region in the north-east of modern Afghanistan. Small chunks of lapis lazuli would have been moved over considerable distances along a route that ran westwards across the northern Iranian Plateau, south of Lake Urmia before crossing through the mountains to Gawra. Copper was also being imported from Iran in limited quantities, while arriving from eastern Anatolia were lumps of obsidian (volcanic glass). As at Susa and Eridu, the community was focused around a series of monumental buildings, a complex of three large tripartite buildings, perhaps temples. Slightly later in date the town comprised a densely packed community with a number of buildings of tripartite plan, possibly the homes of elites who were expanding their authority through the management of resources and their production. They were aided in this work by administrative technology that took the form of stamp seals.

FIG. 9 Painted pottery beaker, ht. 21.8 cm; bowl, diam. 19.5 cm; jar, ht. 6.1 cm, from the cemetery at Susa, Iran, *c.*4200–4100 BC, Louvre Museum loan 1076 © Ashmolean Museum, University of Oxford

**FIG. 10** Stone stamp seals and modern impressions, *c*.4300–3800 BC, max diams. 2.7–4.5 cm, AN1896–1908.O.8, AN1920.68, AN1872.342, AN1912.148 © Ashmolean Museum, University of Oxford

## SEALS AND THE SHAMAN

A useful tool in the organisation of both storage and the control of the extensive exchange networks that had emerged in the fifth millennium BC were stamp seals. Small pieces of stone (one to two centimetres across) were carved onto a variety of shapes, usually pierced with a hole for suspension, with a geometric design cut into the flat base. They may have originated as amulets – and would always retain something of their supernatural qualities – but at sites in Syria by the end of the seventh millennium BC the engraved surface was being impressed on lumps of wet clay or plaster placed on the fastenings of baskets, ceramics, stone vessels, sacks and storeroom doors. In this manner, seals helped to define individual property and secure the containers and rooms against unauthorised opening. The earliest impressions (sealings) may not have been produced by high status members of the community but simply as an effective method of managing the movement and storage of the goods, for example, by herders who would return to the stores after being away from their villages for a certain part of the year.

Seals became part of a system that monitored the production, movement, and distribution of raw materials, foodstuffs and finished goods. By 4200 BC, some sealing activity appears to have been increasingly under the control of more limited groups of people who were restricting access to goods for their

FIG. 11 Stone stamp seal or pendant with 'Shaman', *c.*4300–3800 BC, ht. 3.7 cm, AN1969.360 © Ashmolean Museum, University of Oxford

own social and economic advantage. Social distinctions in the villages and towns started to become more apparent. As with pottery and architecture, there was a widespread sharing of this administrative technology across Mesopotamia and the western Zagros as well as perhaps the symbolism that was expressed in the imagery of the seals. Stamps gradually came to include representations of not only geometric patterns and horned animals but also snakes, vultures, and humans (fig.10). Of particular interest is the appearance of a humanoid figure with the head of an animal (fig.11). Although the identity of this figure remains unknown, it has been suggested that he was a kind of shaman, an individual who is able to interact with the spirit world through an altered state of consciousness. Whatever the correct interpretation, the figure is known from seals and impressions within the foothills and mountains reaching from Degirmentepe in eastern Anatolia (modern Turkey) to Tepe Gawra in northern Mesopotamia, and Tepe Giyan in the western Zagros to Susa. The horned shaman on seal impressions from Susa wears a long skirt and is shown either engaged in activities, perhaps rituals, or, with outstretched arms, grasping snakes and lions (fig.12).

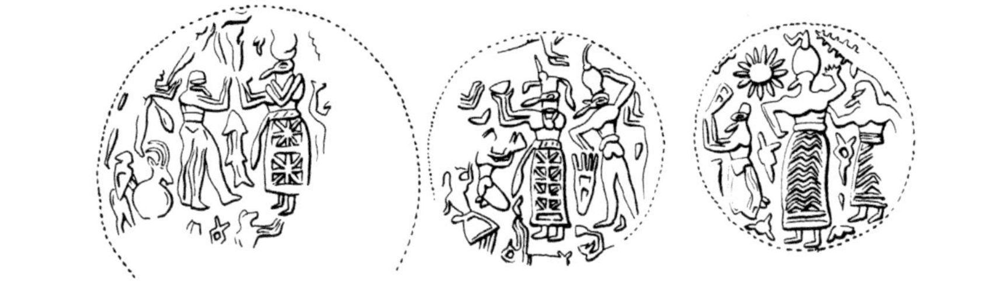

FIG. 12 Drawings of clay seal impressions, from Susa, Iran, depicting the 'Shaman'. After P. Amiet, *La Gyptique Mésopotamienne Archaïque* (Paris, 1961)

## THE METALWORK REVOLUTION AND CHANGING ECONOMIES

If styles of pottery are a guide to the links that existed between Susiana and southern Mesopotamia, by 4200 BC these were growing weaker while, as demonstrated by the ceramics from the Susa cemetery, connections with the Zagros were growing stronger. It has been suggested that there may have been increasing tensions between the two regions over areas of pasturage and access to resources at this time. If this was the case, its impact is likely to have been quite limited as populations would have been relatively small, the pasture lands extensive and the demand for resources limited. Nevertheless, in the period around 4000 BC settlement on the Susiana plains declined, perhaps by as much as two thirds, to leave a population of only about 5,500 people. The town of Susa shrank in size, the buildings on the high platform of the Acropole mound were burnt and the facade of the platform collapsed. Whether all these events were connected is unknown but they may be linked with the onset of a period of severely reduced rainfall over the Persian Gulf so that, for the following 1,000 years, expeditions for resources by boat from Mesopotamia and Susiana were much reduced and overland routes became the main methods for trade and communication.

Of all the resources available in the mountains of Iran the one that would play arguably the most fundamental role within the societies of both the highlands and the lowlands was metal, especially copper. It would be used for weapons, tools, and items of adornment as well as in the service of temples and palaces. Unlike Mesopotamia, which is completely lacking in metal, Iran has plentiful deposits, especially of copper and copper-bearing ores, but also of lead, silver, and gold. It was in the mountains that the technologies of extraction and processing were developed and it gave the highland societies that developed around these activities economic and political influence. Metal became highly prized by the lowlanders who resorted to either negotiation or aggression to acquire it. Metal sources across the plateau had been tapped as early as the eighth millennium BC, but the origin of metal smelting technologies can be identified at sites during the fifth millennium BC, including the earliest alloying of arsenical copper and the eventual adoption of tin-bronze. By 4000 BC there already existed a number of settlements devoted to intensive copper smelting and casting. At sites such as Sialk, for example, copper objects were produced in moulds together with the manufacture of wheel-made pottery, and stamp seals were used to help manage a highly organised approach. During the fourth millennium BC there was a dramatic expansion in metal extraction with the widespread use of copper alloys alongside lead, silver, and gold. Communities in the central and southern Zagros were involved in its local use and distribution but ingots and finished items – simple tools and weapons, and grave goods – travelled along trade routes into Mesopotamia and Susiana.

The social and economic developments of the later fifth millennium BC became even more marked after 4000 BC. At Tepe Gawra, for example, production included cloth manufacture, pottery making, wood working, the carving of bone and stone beads and of seals, as well as the manufacture of ground stone and flint tools, some of which was focused in particular buildings. A large round two-storey building, measuring roughly sixteen metres in diameter, was built at the centre of the town. The thickness of the walls suggests that it served as a fortress for the community as well as a place for storage; a ramp led down to a series of underground rooms, one of which held grain. At a slightly later date, the town was dominated by a number of tripartite houses each with a large central room, some with hearths, flanked by smaller rooms. These buildings, which perhaps served as the houses of the elite, are distinguished by auxiliary rooms that extend beyond the central room to form a deep porch – a feature that may point to links with Anatolia.

Indeed, connections between Gawra and the upland societies of both Anatolia and especially Iran, together with the trade routes that crossed the plateau, are demonstrated by the finds from a number of burials that contained a wealth of objects, especially beads – as many as 25,000 in a single grave. The beads were made from exotic stones: lapis lazuli, quartz, limestone, haematite, carnelian, and turquoise. In addition, small amounts of gold were imported to fashion ornaments in the shape of rings, rosettes, and crescents that were cut from thinly hammered sheet metal and probably then attached to clothing. Other fine and unusual objects included a boar's tusk ivory comb, ceramic ointment vessels, and a tiny wolf's head with movable lower jaw a mere three centimetres in length made of electrum (an alloy of silver and gold). The quantity and quality of these finds suggest that the burials belonged to individuals with a degree of status in a society that demonstrated increasing social stratification.

That there were established connections along the piedmont route to the south of Gawra is suggested by similar forms of tripartite architecture known from the site of Qalinj Agha, close to Erbil, and further south still in the Jebel Hamrin at Abada and Kheit Qasim (see fig.6). All these well-constructed buildings contained the evidence of ordinary domestic use – cooking and eating – but, as at Gawra, they were also places where manufacturing took place and may have been the residences of people with status.

Gawra was also connected with regions further west, across the River Tigris on a route that ran south of the Jebel Sinjar where more tripartite architecture has been identified at the site of Grai Resh. To the north of the Jebel Sinjar, other routes converged on the fertile lands around the tributaries of the Khabur River – the so-called Khabur Triangle in modern Syria – and one of its most important settlements, Tell Brak. The economic and social developments of a market town like Gawra pale in comparison with the developments that were taking place there.

## THE BIRTH OF CITIES

Between 4000 and 3800 BC Tell Brak grew rapidly from some 55 hectares to about 130 hectares, making it appreciably larger than sites like Eridu on the southern alluvium. It came to dominate the neighbouring villages which were eventually abandoned, their populations perhaps emigrating to Brak where monumental mudbrick architecture was constructed. Along with this extraordinary urbanisation came changes in the local economy, including specialisation, with mass production concentrated at the edges of the site. The people who managed these centralised activities had wide-ranging authority over workers and storage, in addition to the distribution and exchange of products. How they gained and held their positions (perhaps through kin relations or political factions) is unknown. That there were tensions created between or even within these communities as a result of such rapid urbanisation may be reflected in the discovery at Tell Brak of a series of mass graves of young adults dating to between 3900 and 3600 BC. These seem to be the outcome of conflict, possibly a civil war.

Tell Brak was not the only urban centre to have emerged on the northern plains but large sites like Hammam et-Turkman and Hamoukar were separated by great distances. It is possible that wild asses had already been domesticated sufficiently for them to be used to transport bulk commodities across the steppe helping to create a culturally connected region – as evidenced by the excavation of small votive plaques known as Eye-Idols at a number of these sites (fig.13). The discovery at Brak of a small clay tablet incised with the image of a wild ass may point to the increasing importance of this animal, as well as representing an early form of notation. Otherwise, human porters were the normal means of moving materials east and west. Whatever the method of transportation, connections were established over prodigious distances and the piedmont route linked the regions of the Khabur River with Susiana.

Urbanism also emerged on the southern alluvium but followed its own trajectory. In this period there was abundant rainfall in the headwaters of the Tigris and Euphrates and the rivers carried much more water than they do today. The two streams merged in the region of modern Baghdad, from where multiple branches created a network of natural channels across the vast alluvial plain. Surface surveys by archaeologists – charting the date and size of settlements based on the distribution of distinctive styles of pottery – have revealed a pattern of dense settlement concentrated in the eastern half of the central plains with several large towns and many smaller villages ranged along branches of the river. As in the north of Mesopotamia, these towns, each of several thousand people, were an attempt to confront the problem of periodic, unpredictable shortages and were concentration points for storage and distribution of food. People were persuaded or compelled to take up residence in these centres. In the southern half of the alluvium, where the landscape was much more waterlogged with extensive marsh lands, settlement took a different pattern, with many smaller sites and only a few substantial centres, the largest of which was Uruk.

FIG. 13 Limestone 'Eye Idols', from Tell Brak, Syria, *c.*3600 BC, ht. 3.2–6.3 cm, AN1939.121–123, 127–129 © Ashmolean Museum, University of Oxford

**FIG. 14** Baked clay bevel-rim bowls from Nineveh and Jemdet Nasr, Iraq, *c.*3200 BC, diam. 17 cm, AN1930.692a, AN1981.986 © Ashmolean Museum, University of Oxford

Undecorated Uruk-style pottery became a feature of these alluvium settlements, both large and small. The most distinctive ceramics produced in vast numbers were, however, bevel (or bevelled)-rim bowls and rounded rim trays (fig.14). At a time when pottery started to be mass produced on simple 'fast' wheels (clay disks turned by an assistant), bevel-rim bowls were made by hand, probably with the help of a mould, before being fired. Both the bowls and trays may have been used in bread making, a staple of the diet; the similar size of the bowls probably reflects a standard 'loaf'. The bevel-rim bowl would come to serve multiple uses as a cheap, disposable container.

This Uruk-style pottery also began to be produced across the Susiana plains where a similar economy based on the herding of cows, along with some sheep, goats, and pigs, and the cultivation of wheat, was now being practised. The upper part of Susiana became densely settled, although this represented only about half the number of people occupying the Mesopotamian alluvium. It is unclear if this transformation represents the immigration of people from Mesopotamia or the natural growth of the local population or a combination of both. By around 3800 BC Susa had once again emerged as an important centre (although its twelve hectares appears small when compared with some sites on the alluvium: Eridu was around forty hectares and Uruk around seventy hectares). The Acropole mound that had been abandoned at the beginning of the fourth millennium BC was reoccupied.

After a couple of centuries, Susa had expanded to cover an area of at least twenty-five hectares and neighbouring Chogha Mish had been resettled. Links to the alluvium of southern Mesopotamia from Susiana included routes through the Deh Luran Plain where bitumen, flint, other stones, and goat hair were being worked and exchanged. There were also, however, continued relations between Susa and the valleys of the high Zagros and perhaps even with the southern Iranian Plateau. The size of Susa's population probably meant that the city was no longer self-sufficient in grain farmed in its immediate vicinities and must have increasingly relied on nearby settlements to meet its needs through systems of taxation or tribute. These smaller villages became linked to a centralised administration and were part of a hierarchy of authority and control – the essential features of a political state.

## BEARS ALONG THE PIEDMONT

Just as the 'shaman' figure had been shared along the piedmont arc from Susa to northern Syria during the fifth millennium BC, other fabulous creatures now followed this path. A particular theme that appears in the imagery of south-western Iran during the second half of the fourth millennium BC is bears acting like humans. From Susa comes a stamp seal impression and five small stone figures depicting such animals. The bears rest on their haunches and hold their front paws up to their mouths, in a gesture almost certainly intended to represent drinking. This is something that bears do naturally, and given a bottle filled with a sweet liquid, the animal will settle back on its rear, wrapping its front paws around the container. The images almost certainly represent a species of small brown bear (*Ursus arctos syriacus*, Persian bears) that can still be found in Iran today, albeit in increasingly fewer numbers, especially in the Zagros. The animals would have been common in the mountainous piedmont arc during antiquity. Bears are solitary animals in the wild and generally are not dangerous unless they feel threatened. They can be trained and performing bears were common in the past. Susa is not the only site to have produced figurines of this type. Among the thousands of beads and small-scale sculptures excavated at Tell Brak and dating to around 3600 BC were many small stone figurines

**FIG. 15** Stone figures of seated bears, from Tell Brak, Syria, *c.*3600 BC, ht. 5.5 and 4.7 cm, AN1939.137–138 © Ashmolean Museum, University of Oxford

of bears (fig.15). Rather than drinking, they crouch on their backsides and rest their front paws on their knees, although one example kneels and wears a skirt. Found along with the figures of bears were limestone and bone stamp seals or amulets depicting a wide range of other animals: lions, gazelles, ibex, monkeys, sheep, rams, bulls and cows, frogs and hares, hedgehogs, foxes, pigs, ducks, fish, and asses.

## THE LATE URUK PERIOD

In southern Mesopotamia the area under cultivation at the head of the Persian Gulf expanded considerably from the mid-fourth millennium BC but this was accompanied by major changes in the settlement patterns. Although the towns of the alluvium continued to flourish there was an abandonment of many of the smaller villages. A similar decline in settlement took place in Susiana as well as in northern Mesopotamia across the plains beyond the Jebel Sinjar. Many of the inhabitants of these regions may have relocated to Uruk which grew to an astonishing 250 hectares (one square mile) with a population estimated to be in the region of 40,000. What compelled or attracted people to the city and its surroundings is unknown. Since our knowledge of these changes is based on surface surveys it is impossible to know the rate at which they occurred and it may have taken a century or more. Indeed, little can be determined about cities on the alluvium other than Uruk, aside from some indication of their scale. They were, however, all within easy reach of each other along a network of natural river branches and artificial canals. This encouraged interaction, exchange, and possibly competition among them. With irrigation, the rich soils were at least twice as productive as the rain-fed agriculture to the north, and bulk quantities of agricultural products could be moved easily by boat between settlements. The rivers also allowed resources from outside the region to be brought in at low costs, transported downstream on boats and rafts. The growth of agricultural productivity transformed society on the alluvium and legions of workers were organised on an almost industrial scale in farming, herding, and the processing of by-products, especially textiles.

The ability of the city leaders to manage people (perhaps forcibly) is reflected in the construction at the centre of Uruk of truly massive buildings using tens of thousands of standardised mudbricks. Many were built with a tripartite plan and, like the earlier temple at Eridu, clearly have their origins in the layout of ordinary houses of the fifth millennium BC. However, whether these buildings served as temples is unclear, although their scale suggests that they held important ceremonial if not ritual functions – the so-called Temple D, for example, had a gross floor area comparable to that of the Parthenon built in Athens around 3,000 years later. Several of these buildings have more unusual layouts: one was almost 50 metres square, with multiple openings around an enormous central space that could have held large gatherings, perhaps for communal feasts. A much smaller building with a tripartite plan and raised on a high mudbrick platform is more securely identified as having had a cultic function. This structure, known as the 'White Temple' for its thickly plastered white walls, contained an altar. It was separated from the monumental halls by a branch of the River Euphrates that flowed through the centre of Uruk.

The significance of many of the large buildings is suggested also by the elaborate mosaic decoration covering some of the walls (see fig.20). This was created using thousands of handmade clay cones, each around fifteen centimetres long. The cones were fired in pottery kilns after which the blunt ends were dipped into different coloured paints. Finally, the sharp ends were pushed into thick plaster covering the mudbrick walls to leave the coloured ends visible and arranged as triangles, lozenges and zig-zags, perhaps copying the designs found on woven reed mats and textiles that have not survived. This not only produced impressive facades but created a protective layer for the brickwork against the extremes of weather.

A large number of exceptional objects were excavated at Uruk, many found together having been deliberately buried – perhaps the result of the ritual clearance of one or more of the great ceremonial structures. Small, finely modelled limestone and clay sculptures of cattle and sheep may represent sacred animals or offerings to the gods. Some of these figures are inlaid with lapis lazuli or have parts of their bodies fashioned from silver,

**FIG. 16** Alabaster sculpture of the 'Priest King', from Uruk, Iraq, *c.*3200–3000 BC, ht. 18 cm, Iraq Museum 51986 © Gianni Dagli Orti/Corbis

materials which would have come from or been carried across the Iranian highlands. There are also images of humans, the most significant of which is the so-called priest-king. He is depicted in sculpture in the round, carved reliefs, and on cylinder seals, distinguished by a distinctive rolled headband, rounded beard (but no moustache) and a knee-length skirt, although he is on occasions also shown nude (fig.16). The priest-king leads rituals, takes part in wild animal hunts, and punishes bound humans. He also appears on the so-called Uruk Vase (fig.17). This remarkable alabaster vessel, just over one metre tall, is carved in shallow relief with a series of encircling registers. The imagery can be understood as a representation of cosmic order which, from bottom to top, ascends from water that sustains cultivated plants and domesticated animals, through a procession of naked (indicating purity or subservience) men carrying vessels loaded with produce, and culminating in the uppermost register where the priest-king makes a presentation, with the help of an intermediary, to a female. She stands, raising her clenched fist, before two poles with loop and streamer. These poles are symbols associated

**FIG. 17** Alabaster vase depicting a ritual procession, from Uruk, Iraq, *c.*3100–3000 BC, ht. 105 cm, Iraq Museum 19606 © 2015. Photo Scala, Florence

with Inana, the patron goddess of Uruk – perhaps it is she or her priestess who is represented here. Unfortunately, much of the image of the priest-king has been lost but fragments of his belt and skirt survive to show that he performed a leading role in the ritual.

Religion and associated rituals probably played an important part in organising and sanctioning daily life but we should be cautious about inferring too much from this evidence. For this period we know virtually nothing about ordinary houses and their inhabitants or indeed about burial practices in general. The focus of archaeologists on temples and other monumental buildings at Uruk means that the evidence from both homes and graves, that might include exotic materials – and thus provide information about relations with the highlands – is so far missing.

## RECORD KEEPING AND THE ORIGINS OF WRITING

Of all the finds at Uruk, the most significant are surely the hundreds of small, sun-dried clay tablets incised with pictographs and symbols that represent the forerunners of true writing. They were part of the technology of record keeping that developed to manage aspects of centralised production and distribution. The tablets, which record economic matters including the management – often in large numbers – of workers, land, animals, and cereals, are a late development in a long tradition of controlling resources. Along with the use of stamp seals, small clay tokens in a variety of shapes, standing for counted commodities such as animals and jars, had been in use for thousands of years in the small farming and herding communities of Mesopotamia. With the growth of large towns and cities, however, the vastly increased scale, variety, and storage and distribution of commodities meant that new forms of accounting and record keeping became necessary. By around 3500 BC, the small tokens were being enclosed in clay envelopes in the shape of balls (the earliest examples are known from Susa and Chogha Mish). The exteriors of the balls (or bullae) were impressed with seals and also tokens, perhaps equivalent to those inside (fig.18). A bulla could be broken open at a later stage as a check on the transaction it recorded and perhaps authorised.

There is good evidence for networks of distribution between the large towns and villages in the form of the stamps used to seal jars, bales and storeroom doors. Now, however, the development of drill technology to produce stone vessels led to new seal designs and the development of cylinder seals in southern Mesopotamia and Susiana. Susa has produced one of the earliest stratified seals of this type. It depicts two rows of fish aligned head to tail and is similar to a type also known from tombs dating to around 3500 BC at the site of Naqada in southern Egypt where they were discovered together with lapis lazuli beads. Both the seals and the beads were clearly imports, small portable objects that were carried easily between towns and villages from Susiana or Mesopotamia along the piedmont before crossing the Syrian plains to reach the eastern Mediterranean and Egypt, fitting into the pattern of long-distance contact we have already encountered.

Around 3300 BC impressions from tokens were being made on flat clay tablets along with a few signs scratched into the surface with a stylus to indicate the objects being counted. A century or so later a highly sophisticated system had been developed that used around 900 signs. Tablets were now divided into grids with each 'cell' containing signs representing ideas and objects alongside numbers organised to some extent like a modern spreadsheet. These documents were used to carry out future checks on, for example, the amounts of barley delivered to a particular granary on a specified day or to keep track of the amounts of barley or beer distributed to workers. It was a complex, time-consuming bureaucratic operation that needed specialised scribes who had been through a period of centralised training. All the texts they produced dealt with local concerns; there does not appear, for example, to be any evidence for the control of long-distance trade. In many respects, the management systems were simply an extension of what had gone before, with landed estates and institutions being organised

**FIG. 18** Clay bulla with seal impressions and tokens, from Susa, Iran, *c.*3300 BC, diam. 7.8 cm, Sb1940 © RMN-Grand Palais (musée du Louvre) / Franck Raux

**FIG. 19** Limestone cylinder seal with silver ram handle, *c.*3200–3000 BC, ht. 5.3 cm, AN1964.744 © Ashmolean Museum, University of Oxford

and administered like large households. After a certain time had elapsed, the information recorded in clay was no longer useful and the unbaked tablets were simply thrown away, landing on refuse heaps with broken bricks and pottery that would later be used as packing for the foundations of some monumental buildings at the heart of Uruk (where they would be uncovered by archaeologists).

Although cylinder seals were rolled across bullae, only rarely were tablets sealed. When this did take place however, seals were rolled over blank tablets, producing something equivalent to a modern piece of headed paper. The information that was then inscribed on the tablet was presumably authorised by the seal impression where the design perhaps indicated a specific department of the central administration rather than the personal identity of an individual. Indeed, many of the earliest cylinder seals were large and almost certainly not designed to be worn. Some had small sculptures of animals carved at the top or dowelled into a perforation running through the cylinder. The magnificent seal in fig.19 is among the finest examples surviving from the late fourth millennium BC. The carved imagery depicts a row of cattle, shown overlapping to indicate a closely packed and moving herd of unlimited numbers, below which appears a line of reed huts containing calves and vessels, the latter presumably for milk. The seal was probably owned by a senior official responsible for the cattle and dairy products of a prominent household or institution. The poles with three pairs of rings that emerge from the huts may represent divine standards suggesting a connection with a temple estate. The top of the cylinder is dowelled to hold a recumbent silver ram as a handle; the use of expensive silver, imported from Iran, helps to emphasise the high status of the seal owner.

A less attractive, yet equally useful, addition to the management tools of these alluvium cities and their institutions were bevel-rim bowls, widely used across the region for baking bread and as an all-purpose container (see fig.14). Its regular size, as well as possibly its relationship with the basic food of life, meant that the vessel came to be used by the central authorities as a measure for rations from their stores: some clay tablets record that labourers were paid in different numbers of bowls' worth of barley or oil.

## SOUTHERNERS MOVE BEYOND THEIR HOMELAND

The development of organised state societies on the plains of southern Mesopotamia and Susiana created new political, economic, and social demands. This led in turn to the establishment of novel relationships with the societies of Syria and upper Mesopotamia. Between 3500 and 3100 BC a number of people from the south followed the piedmont or the Euphrates valley onto the northern steppe lands. This contact with the north took different forms depending on the relationships that were established with the existing local communities. A certain amount of interaction – which was not all contemporaneous – may have been state controlled and sponsored, some perhaps the result of groups fleeing disadvantage at home or following a leader excluded from power. The presence of these southerners took the form of colonies or enclaves within settlements or even perhaps the takeover and occupation of entire sites. They have been identified by the presence of distinct Uruk-style pottery, architecture and seals. At Tell Brak, for example, a temple was built with southern style architecture and forms of decoration including cone mosaic (fig.20). A few sites of pure southern character were

implanted in areas that had not seen previous occupation. One such town, overlooking the Euphrates in Syria, was the site of Habuba Kabira that was built from scratch with standardised houses (using Uruk-style bricks) divided by small alleys leading to main streets that were paved and provided with drains. The settlement was enclosed behind an impressive wall and the highest part of the site crowned with monumental buildings just like those at Uruk. The level of planning and organisation that would have been needed to create Habuba is astonishing. The site may perhaps have coordinated the activities of certain enclaves of southerners elsewhere in the region, including the processing of local products such as wool and flint as well as the exchange of finished goods, including perhaps textiles imported from the south, for other resources.

A number of the people moving to these northern 'colonies' may have originated in Susiana as the region witnessed a decline in settlement and a loss of populations. There were also significant movements of people within Susiana itself: Susa shrank from an area of twenty-four to only nine hectares while Chogha Mish grew from eight to nearly eighteen hectares. By around 3300 BC, the two sites were separated by a fifteen-kilometre-wide stretch of territory devoid of villages, which suggests there had been a breakdown in relations and perhaps conflict between these two political centres. The inhabitants of the abandoned villages may have migrated to Chogha Mish. Others may have moved to the immense cities in southern Mesopotamia, or perhaps further afield to the settlements being established at locations in the north and, as we will see below, at Godin Tepe in western Iran. Although Susiana shared a common culture with the cities of southern Mesopotamia, it is probably unlikely that there were attempts at political domination by cities like Uruk. If this had been the case it might be expected that all of the Uruk systems of administration would have been introduced to the region but, of the thirteen notational systems in use at Uruk, only three have been documented in the administrative texts found at Susa.

A lack of archaeological work in Iran that explores this crucial period makes it difficult to say a great deal about relationships between the lowlands and the highlands. The stamp seals and their impressions from Tepe Giyan show no influences from the alluvium but are very much part of local Zagros traditions that can be traced across the mountains and on into the north-east piedmont of northern Mesopotamia and places like Tepe Gawra. Bevel-rim bowls start to be produced at a number of sites across the Iranian Plateau but this does not necessarily indicate the presence of people from Mesopotamia but rather the adaptation of similar methods of food preparation, perhaps associated with the emergence of more complex

**FIG. 20** Section of baked clay cone mosaic, from Tell Brak, Syria, *c*.3200–3100 BC, ht. 43 cm, AN1939.330 © Ashmolean Museum, University of Oxford levels of social organisation. However, one site, Godin Tepe in the western Zagros, does suggest connections with the lowlands were established on a more permanent basis. Godin Tepe is positioned strategically at the centre of the Kangavar Valley and it must have benefited from, if not controlled, the flow of materials and products along this important route. Sometime around 3200–3100 BC an unusual mudbrick oval enclosure was constructed around a thick-walled building at the highest point of the settlement. Although the building technique was in a local highland tradition, the people living within the enclosure used some Uruk-style pottery (although at least 50 per cent was in a local style) along with cylinder seals and numerical tablets that are closely comparable to those known from Susa. If, as has been suggested, the oval acted as a kind of fort, it is possible that the residents were from the lowlands, possibly Susiana, and worked with or even dominated the local population.

The central authorities in the lowlands would have coveted the resources of the Iranian highlands but may have had to acquire them through middlemen rather than directly, as appears to have been the case on the steppe lands of northern Mesopotamia. Whether the Iranian mountains were a source of good quality timber for roofing, which would certainly have been necessary to span colossal buildings like those at the centre of Uruk, is uncertain as all the later evidence points to its acquisition from areas to the north-west, in Syria and Anatolia. It seems unlikely, however, that the oak, pine, and juniper that grew in the Zagros was not exploited and this timber may have been acquired through a network of relations established with local communities. Along the same routes may have travelled small quantities of exotic stones and the agricultural produce of the fertile mountain slopes and valleys, including wine, which evidence indicates was being produced at sites like Godin Tepe (and consumed in the cities of southern Mesopotamia). Although the evidence from the alluvium is scanty, another resource almost certainly being sought by the lowland cities was metals, especially copper. During the fourth millennium BC copper ores were being mined on the Iranian Plateau in large-scale operations. At the site of Arisman, for example, people lived in a large settlement close to the mines where they extracted the ore and then smelted it nearby. It is likely that much of this metal was used and distributed locally for tools, weapons, and other objects of status – there is certainly no evidence for populations with an Uruk-style culture being involved with its production. Greater demands, however, would have encouraged contact and expansion of production and the people of Mesopotamia now tapped into the supplies. Ingots of copper, as well as more exotic metals would have been carried along the main Zagros arteries, by donkeys or human porters, through settlements like Godin Tepe. Indeed,

**FIG. 21** 'Skidding goat' pottery vessel, *c.*3200 BC, ht. 12.5 cm © Ashmolean Museum, University of Oxford

the increased number of settlements that developed at this time along the Zagros valleys of Mahidasht and Kangavar highlight the importance of this route. Here a distinctive painted pottery was produced; most of the motifs decorating the exterior of these vessels are geometric but rare examples include leopards, birds, and 'skidding' goats (fig.21). At the western end of the route lay the crossroads giving access to the north–south piedmont section of the network, and villages like Tell Rubeidheh in the Hamrin Valley where a mixture of pottery typical of both highland and lowland was in use.

## FANTASTIC ANIMALS FROM IRAN TO EGYPT

As we have traced, the system of highland–lowland trade that extended along the flanks of the Zagros and Taurus Mountains and onto the neighbouring plains was already ancient by the end of the fourth millennium BC. The piedmont route connected southern Mesopotamia and Susiana via the foothills of the Zagros with northern Syria through the exchange of materials and products as well as the sharing of images and ideas such as the 'shaman', bears acting as humans, and animal 'amulets'. Clay impressions that sealed containers or administrative documents as well as actual cylinder seals

**FIG. 22** Green jasper cylinder seal and modern impression with serpopards and lion-headed birds, *c.*3200—3000 BC, ht. 4.6 cm, MNB1167 © RMN-Grand Palais (musée du Louvre) / Franck Raux

occasionally also moved over great distances together with the commodities they marked. In this way a distinct range of images of composite, fantastic animals crossed cultural boundaries in the later fourth millennium BC. These 'monsters' were rooted in the tradition of the composite 'shaman' and mixed different animals into a single creature. One of the earliest to emerge was a feline with the wings of a bird known from seal impressions from Susiana. Slightly later in date, a related – if not the same – creature appears in the impressions at Uruk in the form of a lion-headed bird. Other fantastic beings represented in sealings from Uruk include a frog-like monster shown frontally in association with interlaced snakes. Perhaps the most intriguing of these fabulous animals, known widely from impressions at both Susa and Uruk, as well as Habuba Kabira and Tell Brak in Syria, and Tepe Sofalin in Iran, are felines with long, often intertwined necks. These are described as 'serpopards', since they seem to be a combination of a serpent and leopard. Only rarely do the actual seals survive from which such impressions were made, but one magnificent example is carved with lion-headed birds and serpopards (fig.22). Unfortunately, this cylinder seal was not excavated and we do not know where it was found.

Remarkably serpopards also appear on a number of high status objects from the Nile Valley associated with the emergence of a unified state in Egypt around 3200–3100 BC. Among them are stone palettes, originally intended to provide a flat surface on which body pigments were ground, that were transformed into ritual objects with images associated with kingship carved in shallow relief on both sides. The most famous examples were excavated at the site of Hierakonpolis in southern Egypt and include the so-called Narmer Palette (now in the Egyptian Museum in Cairo) and the Two-Dog Palette (fig.23). Both objects are decorated with a pair of serpopards, their long necks flanking the circular grinding area. Those on the Two-Dog Palette lick the body of a gazelle. On the reverse of this palette a mixture of fabulous and real wild creatures occupy the left-hand side of the scene from where they attack or harry a range of herbivores native to North Africa. From the top, there are a pair of lions, a serpopard, a leopard, a hyena, and a griffin with comb-like wings. At the bottom is a long-tailed dog-like creature wearing a belt or penis-sheath and playing an end-blown flute. Although sometimes described as a man wearing an animal mask, this is perhaps better understood as the import of an idea from Iran where, as we will discover, there was a tradition of depicting animals acting as humans. Presumably the imagery was carried to Egypt along the trade routes on portable cylinder seals or their impressions and from these the exotic creatures were adopted by Egypt's early kings to express their otherworldly status and power.

**FIG. 23** Siltstone 'Two-dog' palette, from Hierakonpolis, Egypt, *c.*3200–3100 BC, ht. 42.5 cm, AN1896–1908 E.3924 © Ashmolean Museum, University of Oxford

## HIGHLANDS AND LOWLANDS DIVERGE

Towards the end of the fourth millennium BC people were yet again on the move. For unknown reasons, but perhaps connected to changes in the climate, some groups and perhaps entire communities, began to migrate south from the region of modern day Georgia and Azerbaijan into eastern Anatolia. They brought with them a way of life and a distinctive type of pottery known as the Early Trans-Caucasian culture. By the early third millennium BC this culture had spread over an enormous area reaching from the Caucasus Mountains into north-western Iran, and through the highlands of Anatolia westwards to beyond the headwaters of the Euphrates. This was a world of dispersed villages with no evidence of public buildings or indeed of administrative tools such as seals. These developments are probably to be connected with a contemporary decline in settlement across the northern plains of Syria and Mesopotamia. Around 3100 BC places such as Habuba Kabira and the other enclaves of southerners were being abandoned. These communities were perhaps unable to function effectively in an increasingly rural landscape of villages, their access to resources gradually limited by the breakdown in the previously integrated economies of the highlands and lowlands.

This had an impact across the wider region, not least along the piedmont route. Susiana began to develop closer cultural and perhaps economic and political links with the highlands. While there had been general stylistic features in cylinder seal designs in common between southern Mesopotamia and those used in Susa, these started to take on more regional characteristics that bore more of a resemblance to the stamp seal imagery of the Zagros Mountains where horned animals, felines, snakes, and the horned demon were at home. A distinct material culture spread across Iran, no doubt following exchange and communication routes that had long been in existence but now became more closely integrated. Thus, in the period approximately 3100–2900 BC many of the small towns of the plateau shared the use of specific types of pottery, including bevel-rim bowls, styles of seals, and written tablets. This is known today as the Proto-Elamite culture.

The idea of writing on clay tablets had of course been shared with Susiana from southern Mesopotamia, but in the Proto-Elamite period a completely new system was independently invented (although the earlier numerical signs were retained). This was Iran's first indigenous writing system and may have been developed in Susa – where around 1,550 tablets have been found – and then extended across a network of Iranian towns. It was once thought that this writing might be connected with later

FIG. 24 Heulandite cylinder seal and modern impression, from Susa, Iran, *c.*3100–2900 BC, ht. 3.4 cm, Sb2675 © RMN-Grand Palais (musée du Louvre) / Thierry Ollivier

inscriptions that recorded the Elamite language but Proto-Elamite remains un-deciphered and a relationship between the two cannot be established.

Some of the Proto-Elamite signs were also incorporated into the designs on cylinder seals and probably held great symbolic meaning; it is likely that the sites where they were in use were linked through ritual or high status family relations as much as through trade. The crosses carved on an extremely fine seal from Susa (fig.24) are examples of such signs, as indeed are both the tree and mountain elements that are flanked by mountain goats. Unlike in Mesopotamia, humans were never depicted on Proto-Elamite seals. Instead the Iranian tradition of animals with human attributes is used. The most powerful of these human-like animals are lions and bulls (the most dangerous wild animals of the region). They are depicted standing on their hind legs on a seal impression rolled twice across the width of one of the largest surviving Proto-Elamite tablets (fig.25): a bull symmetrically restrains two seated lions, alternating with a lion dominating two rearing bulls, each topped with a fringed or 'hairy' triangle (another sign found in the Proto-Elamite script). Perhaps the most astonishing portrayal of this type of animal is a stone sculpture of a lioness (fig.26). She appears monumental in scale with a powerfully twisted body but in reality is only eight centimetres high. Sadly, the object was not excavated and the site from which it comes is unknown.

The Proto-Elamite texts relate only to local book-keeping operations and do not suggest any centralised economic or political control. The earliest tablets outside of Susa come from Tal-e Ghazir, Tepe Sialk, Tepe Ozbaki, and Tepe Sofalin pointing to connections between Susiana and the northern-central Iranian Plateau. Slightly later in date, tablets found at Tal-e Malyan, Tepe Yahya, Shahr-e Sokhta, and Tepe Sofalin, link Susa to the central and southern Zagros and eastern Iran. Here the most important settlement was undoubtedly Tal-e Malyan on the Marvdasht Plain which had expanded to around fifty hectares, about five times the size of Susa. A large, elaborately decorated building as well as a warehouse with evidence of craft production show that it was an important political and economic centre that dealt with huge numbers of sheep and goats.

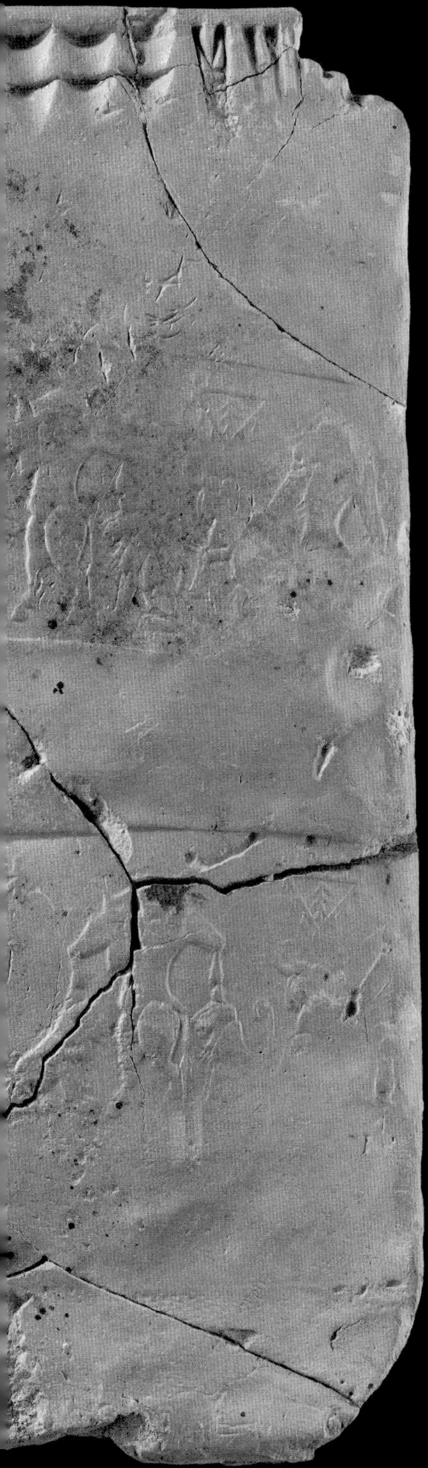

LEFT:

**FIG. 25** Inscribed and sealed Proto-Elamite clay tablet, from Susa, Iran, *c.*3100–2900 BC, ht. 21 cm, Sb2801 © RMN-Grand Palais (musée du Louvre) / Franck Raux

BELOW:

**FIG. 26** Crystalline limestone or magnesite figure of a lioness-demon, *c.*3100–2900 BC, ht. 8.8 cm, Private Collection © Sotheby's

# III. From City to Kingdom: 3000–1500 BC

Limestone votive statuette of a standing male, from Istabalat, Iraq, *c.*2500–2350 BC, fig.34

# III. From City to Kingdom: 3000–1500 BC

THE ABANDONMENT of Uruk-style settlements in northern Mesopotamia and Syria may have contributed to the on-going expansion of the cities on the alluvium if their populations had returned southwards. This process of urban growth was intensified by the desertion of large numbers of the smaller villages across the alluvium by people migrating to cities; Uruk expanded to cover an extraordinary four square kilometres (one-and-a-half square miles). The monumental buildings at the heart of the city were razed to create a series of large terraces to support new structures; the surface levelled by filling holes and pits with debris from rubbish dumps (broken pottery, sealings, and tablets) that had been created by the activities of the central storage facilities.

The city officials continued to use cylinder seals with designs that showed considerable continuity with earlier imagery, including representations of the priest-king. The technology of writing was refined further with signs being much more clearly organised on tablets and their forms becoming less pictographic and more wedge-like or 'cuneiform'. By this date, cities across the alluvium had adopted writing as an administrative tool and similar tablets have been found beyond Uruk at Ur, Umma, Nippur, Kish, Jemdet Nasr, and Tell Uqair, as well as in the lower Diyala at Tell Asmar. A good example of one of these tablets is often referred to as the 'field of the EN' text (fig.27). It contains calculations for the areas of six large fields. The scribe added these figures together and divided the grand total into thirds. Two whole thirds were allocated to a person designated by a sign read as EN, while the remaining third was divided unevenly amongst five other individuals. One of these is the EN.SAL, who some have interpreted as 'wife of the EN'. It clearly reflects a stratified society with EN as a leading figure, perhaps the 'priest-king'. Other texts from Uruk and Jemdet Nasr contain the term NIM.KI that in one instance is used to describe a number of female slaves. Although it is not possible to be certain at this early period, later texts (dating to around 2600 BC, when scribes are recording the Sumerian language) use the sign NIM to denote Elam, a region of south-western Iran. The acquisition of slaves was one of the aims of raids into the Zagros by later Mesopotamian rulers and it is possible the practice was already established at this time. Another sign found in these early tablets is KUR. It is used to distinguish different types of

**FIG. 27** 'Field of the EN' clay tablet, from Jemdet Nasr, Iraq, *c.*3100–2900 BC, ht. 43 cm, AN1926.583 © Ashmolean Museum, University of Oxford

**FIG. 28** Jemdet Nasr-style pottery, from Jemdet Nasr, Iraq, *c.*3100–2900 BC, ht. 25.5 and 12.7 cm, AN1926.496 and AN1925.371 © Ashmolean Museum, University of Oxford

animals, products, and geographical areas. In later Sumerian texts KUR means 'mountain' or 'foreigner' and so, if it held a similar meaning in the late fourth millennium BC, the sign could be indicating imports from the east.

A new feature of the material culture of southern Mesopotamia at this time was the production of a 'Jemdet Nasr' style of pottery painted with geometric and figurative designs in the colours black and plum-red that was derived from haematite (fig.28). This may have been produced for high-status individuals or for specific activities, as other simpler wheel-made pottery was more widely distributed. Examples of Jemdet Nasr ware have been found in so-called Hafit tombs in the United Arab Emirates. It has been suggested that such vessels travelled with Mesopotamian traders interested in acquiring copper from the Oman Peninsula. This reflects a reorientation of trade following the abandonment of the Uruk-style colonies and the revival of maritime links that had been limited since the Ubaid period.

While the cities of southern Mesopotamia were increasingly establishing links along the Persian Gulf in search of resources, the piedmont route remained closely tied to the communities and traditions of the Zagros. A characteristic of this new orientation is the so-called 'Piedmont' or glazed steatite cylinder seal. These seals, first manufactured in Proto-Elamite Iran, were produced over a wide area of the Zagros foothills and adjoining plains including western Iran, the Diyala region, the upper Tigris, and the Sinjar and Khabur regions (map 3, below). Generally decorated with geometric motifs, the seals are covered in patterns such as rosettes, centre-dot circles, and hatched bands of arches or lozenges, the combination of which may have represented messages about the user or the sealed product. One such sealing, perhaps from a jar, was found at Jemdet Nasr (fig.29). The typical curvilinear motifs and crosses of Piedmont seals are here combined with a schematic mountain goat or ibex.

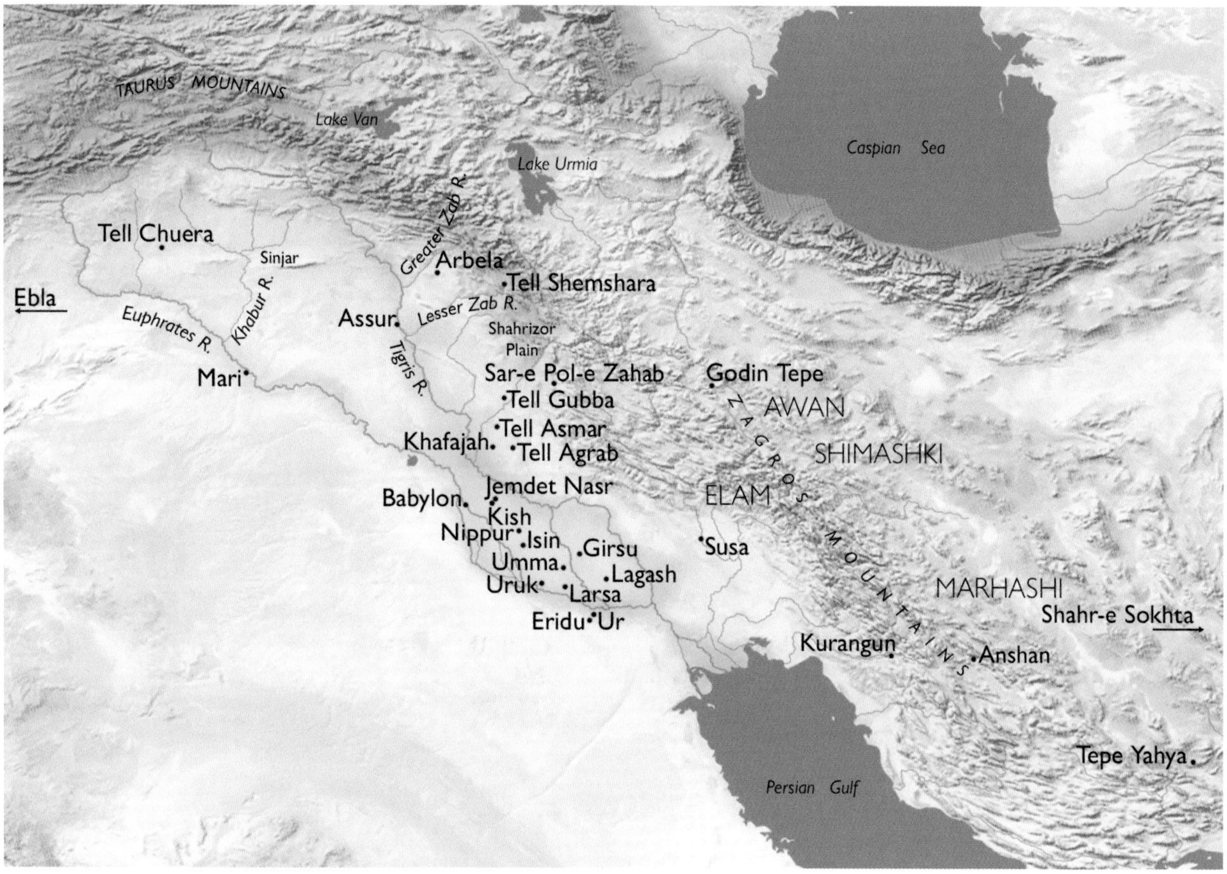

**FIG. 29** Clay sealing with impression of Piedmont seals, from Jemdet Nasr, Iraq, *c.*3000–2800 BC, l. 6 cm, AN1926.678A © Ashmolean Museum, University of Oxford

## ALONG THE PIEDMONT

The residents of the towns and the villages of the Diyala and Hamrin valleys received cultural influences from both the alluvium and the highlands. Pottery used in this region included both painted Jemdet Nasr styles and the characteristic ceramics of the Diyala known as Scarlet Ware, while metalwork showed influences from Iran. In the Hamrin valley a particularly distinctive form of circular building was constructed, of which the best known is Tell Gubba. It consists of a unit formed from a core of masonry surrounded by five concentric walls creating four concentric passageways. The entire structure is around twenty metres in diameter. Like the much earlier Round House at Tepe Gawra, the living area in the Tell Gubba buildings was probably on a higher floor – reached by stairs built into the thick corridor walls. The Hamrin is more suitable for herding than for the cultivation of cereals, and the circular buildings may have served as the headquarters of semi-nomadic groups who made their living from large flocks of sheep and goats. The buildings would have served as depots for food to be used in times of shortage and as defensible bases against marauders, as well as homes for the local chiefs. These buildings may also have represented fortified outposts from the political centres located in the towns of the lower Diyala.

There were probably no similar political centres in northern Mesopotamia in this period but rather small-scale settlements that used common styles of pottery (known as Ninevite 5) in the geographical area from the Khabur to the foothills of the Zagros. Although writing was not used here, cylinder seals carved with Iranian motifs may have been used to distinguish individuals of higher status. Indeed, a certain degree of social stratification is evident in burials that range from modest pit graves to the mudbrick tombs of the relatively wealthy containing numerous copper-alloy pins and beads, some of lapis lazuli.

## THE EARLY DYNASTIC PERIOD

Proto-Elamite writing may have been in use for as little as a century before it ceased to have a practical purpose and was abandoned. Unlike in southern Mesopotamia, writing had not been embedded in society through a centralised system of education and the loss of influence by the small group of people who used it means it soon fell out of use. The disappearance of writing in Iran may have been connected with the decline of the Proto-Elamite culture. By around 2900 BC Proto-Elamite sites of northern-central Iran were replaced by the Early Trans-Caucasian culture. Further south, the entire Zagros region from Godin Tepe to Tal-e Malyan as well as the lowlands of Susiana experienced a decline in the size of settlements. A five-kilometre-long mudbrick wall on a stone foundation and at least two parallel inner walls were built to encircle Tal-e Malyan. While the walls suggest troubled times they also indicate that the city retained the capacity to organise this substantial work. The changes in regional settlement patterns may have been due to the emergence of economies based upon seasonal mobile pastoralism between the highlands of south-western Iran and the lowlands of Susiana.

In southern Mesopotamia, however, urbanism was by now firmly established and many cities shared a similar layout. They became much more densely occupied with temples located in specific areas of the settlement, often at the edge of the city on their own mounds. Uruk at around 400 hectares remained by far the largest centre and was now encircled by a

OPPOSITE, LEFT:
**FIG. 30** Limestone vase or vessel support with heroes and animals, *c.*2800–2600 BC, ht. 12.7 cm, 1927,0409.1 © Trustees of the British Museum

OPPOSITE, RIGHT:
**FIG. 31** Gypsum vase or vessel support with hero and animals, from the Shara Temple, Tell Agrab, Iraq, *c.*2800–2600 BC, ht. 15.2 cm, A17948 © Courtesy of the Oriental Institute of the University of Chicago

nine-kilometre-long mudbrick wall – the building of which reflects an organisation of labour on a truly monumental scale. A small group of clay sealings, known as City Seals, may provide clues to the political, economic, or social relationships across the wider alluvial plain at this time. Some impressions from Ur (with earlier examples from Jemdet Nasr) have a variety of combinations of pictographs standing for the names of some significant cities, including Eridu, Ur, Larsa, Uruk, Adab, and Nippur. These groupings perhaps represent networks of city-states linked by the river branches and canals: the seals may have authenticated receipt of high-status food offerings made to the cults of some of the most important gods.

The administration of the large institutions in these cities continued to rely on scribes trained in the systems of accounting that had their origins in the fourth millennium BC. Although based partially on ideograms, the earlier script did not reflect its users' language. Judging from the content of some tablets from Ur dating to around 2800 BC, however, there was a development in linguistic content. Thus the cuneiform writing system – that would be used for the next three millennia in Mesopotamia – emerged with documents containing expressions reflecting the grammar and syntax of the Sumerian language. There are indications from these early texts that Mesopotamia was a multilingual society and other languages, including one termed by modern scholars as Akkadian (a Semitic language, with a similar grammatical structure to Arabic and Hebrew), were being spoken. Sumerian was likely a language of the elite that scribes would have had to learn to write even when it was not the language they spoke on a daily basis.

Much of our evidence for relations between Mesopotamia and Iran in the early third millennium BC comes from excavations in the larger towns of the lower Diyala valley. Their location on a north-eastern extension of the alluvium, close to the piedmont and a principal route onto the Iranian Plateau, resulted in a fusion of ideas and traditions. As we have seen, the Diyala settlements produced a distinctive style of pottery known as Scarlet Ware. The decoration of these ceramics became more elaborate over time with representations in red outlined in black of humans, and animals such as birds perched on the back of goats. Similar imagery is also known from stone sculpture that was dedicated in temples, including an elaborately carved vessel or support for a round bottomed jar which is carved with a large bird perching on the back of a pair of bulls (fig.30). Two bearded men, nude except for double-strand belts, hold the bulls, one with his arms around the animals' necks while the other grasps their rumps. This sculpture is unprovenanced but a strikingly similar vessel was excavated in the so-called Shara Temple at Tell Agrab in the Diyala (fig.31). Here our 'hero' grasps

the rumps of two lions while the tails of two other lions are held under his arms. All four felines menace a bearded bull or bison facing away from the man. Around the neck of the vessel and under the feet of the highest lions is carved a scale-like design which is a standard Mesopotamian artistic motif for depicting mountains, presumably the Zagros. The meaning of the imagery is unknown but a hero figure confronting animals is also known from cylinder seals of this period (fig.32) and later (see fig.42), and associates the eastern mountains with wild creatures and a mythological realm where heroic figures confront and master them. Interestingly, the stance of the heroes, with their upper body shown frontally but twisted to the right at the waist and legs, is identical to that of the lioness demon in fig.26, suggesting links with the Proto-Elamite past.

**FIG. 32** Brown marble cylinder seal and modern impression with hero and wild animals, from Kish, Iraq, *c.*2700–2600 BC, ht. 2.8 cm, AN1925.95 © Ashmolean Museum, University of Oxford

**FIG. 33** Alabaster plaque with scenes of banquet and animal combat, from Susa, Iran, *c.*2700–2600 BC, ht. 17 cm, Sb41 © RMN-Grand Palais (musée du Louvre) / Jérôme Galland

Other objects dedicated in the temples of the Diyala and southern Mesopotamia were flat, square stone plaques with a hole through the centre. They were carved in relief on one side into a series of registers in which people are shown in the lower sections travelling by foot, boat, or chariot and often accompanied by animals, towards elite or divine banquets at the top. Using a peg driven through the central hole, the plaques were fixed to a temple wall beside an entrance so that a rope attached to the door could be wound around the head of the peg as a locking mechanism. Stone plaques of this type were a tradition that was shared with Susiana. This could result in local understandings of the imagery so that in a plaque excavated at Susa, the lower section depicts not a procession but a man using a spear to ward off a lion that attacks a cow or bull (fig.33). In the upper half a seated woman plays a musical instrument – in Mesopotamia she would be shown holding a bunch of dates while a servant played the harp or lyre.

Lockable doors and small cult rooms in Mesopotamian temples strongly suggest that they were not intended as congregational spaces and that access was limited and probably restricted to the priests. It may have been because of this that the elite commissioned images of themselves to be carried into the god's presence. These stone statues embodied the essence of the worshipper so that the spirit would be present while the physical body was absent. All such statues have been discovered buried in groups under the temple floor, or built into cultic installations such as altars, or scattered in pieces in the shrine and surrounding rooms. As a result it is unknown how, or if, the statues were presented to the god. The statues depict both men and women wearing fringed or tufted sheep fleece robes – a ritual garment perhaps intended to associate the wearer with the animal most regularly sacrificed to the gods (fig.34). Figures either stand or sit, often holding their hands clasped at the chest or waist. Others hold cups or branches of vegetation, such as a seated female excavated at Kish (fig.35). A large number of votive statues dating from around 2900 BC were discovered in temples at Tell Asmar, Khafajah, and Tell Agrab in the Diyala. Later examples are known from sites as far afield as Nippur, Assur, Tell Chuera, Mari, and Susa (fig.36).

**FIG. 34** Limestone votive statuette of a standing male, from Istabalat, Iraq, *c.*2500–2350 BC, ht. 20 cm, AN1919.65 © Ashmolean Museum, University of Oxford

LEFT:
**FIG. 35** Limestone votive statuette of a seated woman, from Kish, Iraq, *c.*2600–2350 BC, ht. 12 cm, AN1924.258 © Ashmolean Museum, University of Oxford

RIGHT:
**FIG. 36** Alabaster votive statuette of a standing man, from Susa, Iran, *c.*2900–2350 BC, ht. 14.8 cm, Sb77 © RMN-Grand Palais (musée du Louvre) / Béatrice Hatala / Konstantinos Ignatiadis

Although the votive objects were almost certainly commissioned by people with wealth and status, few are recognisable from their images as holding positions equivalent to a king or queen (by this time the image of the priest-king had fallen out of use). There are, however, some indications of the existence of local rulers from fragmentary inscriptions but their authority and roles may have varied greatly between cities. On the whole, society in southern Mesopotamia seems to have been organised according to households, ranging from small domestic family groups to substantial dynasties, together with their property, including fields, animals, and slaves. The largest households, at least across the lower alluvium, were probably the temples, sometimes with hundreds of dependants including farmers, herders, and textile weavers, and these played a vital role in the political and economic organisation of their towns and cities. In the northern half of the plain, however, a regional power emerged around 2700 BC centred at the city of Kish where some of the rulers styled themselves as LUGAL, literally 'big man' but generally translated as 'king'.

At Kish an impressive mudbrick palace was built consisting of several buildings ranged around a large open court: one building was approached by a majestic staircase; a four-columned portico formed the facade of another. This is the first such building that can be identified with a degree of certainty as a palace – a monumental, non-religious building with reception areas and associated workshops. Some of the internal walls of the columned building were decorated with impressive imagery (fig.37). Shallow depressions were hollowed out of slate to hold limestone inlays. A bearded soldier wearing a flared hat and rolled-up skirt is shown walking to the right with the outline of a companion at his rear; the vertical strip of inlay is probably a border dividing the scene. The soldiers were originally shown escorting bound and naked captives towards the figure of a seated ruler (that part of the inlay is in the Iraq Museum, Baghdad). Warfare between city states and military raids is thus emphasised as an important royal activity. This image of military activity, or at least its aftermath, is among the earliest examples of one of the two recurring themes of art in which the king participates, the other being banquets celebrating the abundance of the agricultural world.

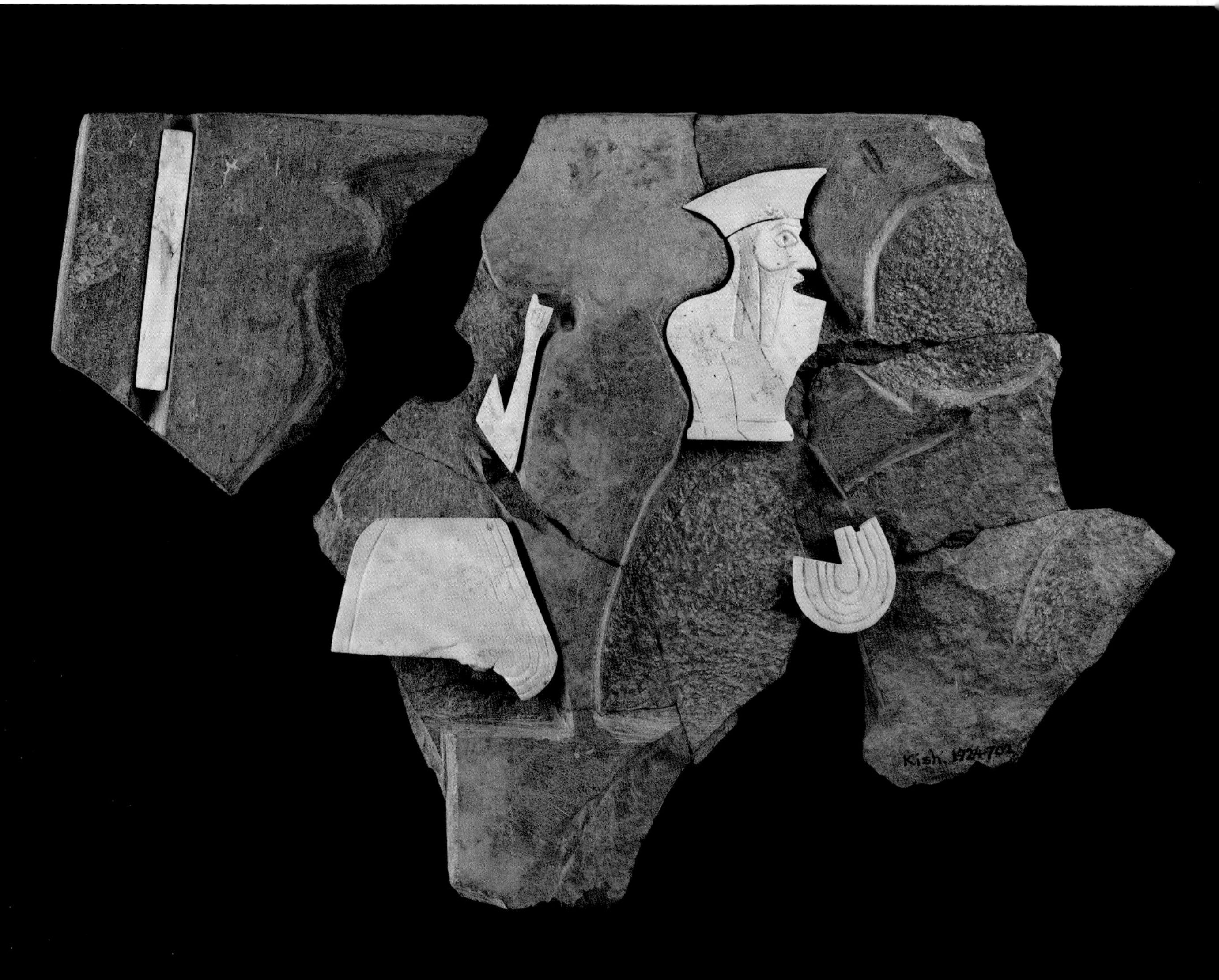

**FIG. 37** Slate and limestone wall inlay showing a procession, from Kish, Iraq, *c.*2600–2500 BC, W. 27 cm, AN1924.702 © Ashmolean Museum, University of Oxford

From the middle of the third millennium BC royal inscriptions include references to attacks by Mesopotamian city-state rulers on a land called Elam. While Sumerian texts refer to a land termed NIM (which means 'high' or 'elevated'), in later texts the sign is equated with the Semitic Akkadian term *elamtu*, perhaps related to the word *elûm* ('high, upper'). Despite these names focusing on high mountains, the region of Elam – although not always politically unified – combined the Susiana plains and the mountains of the southern Zagros. The raids into Elam were focused on acquiring the resources of the region including, and perhaps especially, the acquisition of slaves. In cuneiform texts male slaves are designated as NITA.KUR and female slaves SAL.KUR, that is foreign/mountain male and foreign/mountain female.

Some of the earliest references in royal inscriptions to Elam come from the monuments of Eannatum, a ruler of the state of Lagash (about 2450 BC) in the eastern alluvium. His campaigns towards the mountains were directed through a place called Arawa and were perhaps attempts to gain access to the lucrative plains around Susa:

> *Elam, 'the high mountain', was smitten with weapons; mounds of corpses were piled high; Arawa, whose governor had raised its standard, was smitten with weapons; mounds of corpses were piled up.*

Eannatum, or perhaps one of his successors, seems to have been successful in establishing control over the routes into Susiana as later texts from Lagash refer to considerable trading activities between that state and Elam: boats from Elam are listed as part of a trading expedition; barley is traded for wool from Elam; silver is delivered from Elam for the purchase of goods; livestock is purchased in Elam; and wooden objects are brought by boat from Elam.

Some of the long-distance routes that had once crossed Iran were gradually reoriented towards the maritime trade of the Persian Gulf. Susa probably continued to play a key role in these connections and after about 2500 BC the city expanded from about ten hectares to upwards of forty-five hectares with many small satellite villages. There is little evidence, however, to suggest that it played a significant political role. Other large centres also emerged at this time on the Iranian Plateau, many in areas, such as in the south-east, which had not previously witnessed noticeable urbanisation. Shahr-e Sokhta, for example, expanded considerably to around 100 hectares along with numerous smaller neighbouring sites. The settlement may have played a role as a trading and manufacturing centre. A workshop for lapis lazuli discovered here may have been supplying the stone for local, as well

**FIG. 38** Beads of carnelian, lapis lazuli, sheet gold, silver, and rock crystal, from Kish, Iraq, *c.*2500–2350 BC, modern restringing l. 40 cm, AN1925.261 © Ashmolean Museum, University of Oxford

as more distant, consumption: hundreds of burials in a cemetery at the site contained worked stones (lapis, turquoise, carnelian, alabaster) along with goods from every culture in the Indo-Iranian borderlands.

Although overland routes would have supplied Mesopotamia with some exotic stones much more would have been transported by sea. Such wide-ranging connections are illustrated by beads found in graves at sites like Kish (fig.38). Long red-orange carnelian beads were a type made by craftsmen in the Harappan Civilisation of the Indus Valley, although some were probably made in Mesopotamia by migrant craftsmen from the Indus who used their own style of drills and raw materials. Some carnelian beads were decorated with white designs. This tradition also originated in the Indus Valley where beads were painted with soda and heated to bleach the stone (they are described erroneously as 'etched'); the technique was adopted and developed in other places including Mesopotamia. Carnelian beads are frequently found in Mesopotamia together with blue lapis lazuli beads, a combination that may have had symbolic and magical properties.

## MYTHOLOGICAL WORLDS

On rare occasions, Mesopotamian graves also included vessels carved in elaborate relief decoration made of soft stones such as chlorite and steatite. Further examples have been excavated from Mesopotamian temples and palaces, while the patterns and figures carved on the containers are known from objects found over a wide area extending from the Persian Gulf and western-Central Asia to the Indus Valley. Many of the designs on the vessels, as well as the stones from which they are made, originated in Iran. One source of chlorite, as well as a production centre for stone containers, was at Tepe Yahya at the southern end of the Zagros chain. The chlorite vessels were probably for local use although what this was is unknown. A few of the containers joined the movements of exotic stones and metals in all directions across the plateau. Some are carved with highly complex designs that include humans and animals and incorporate Mesopotamian, Iranian, and Harappan imagery showing the wide-ranging interconnections that existed in the second half of the third millennium BC. One particularly fine example includes a central male figure, repeated twice, with long plaited hair and wearing a skirt decorated with incised cross-hatching (fig.39). One man stands on the rumps of two lions positioned back-to-back and clutches a pair of large serpents with gaping mouths and spotted bodies. The other male, grasping streams of water rather than snakes, sits on the rumps of two humped bulls or zebu (animals found in the imagery of the Indus Valley civilisation). A third section of the vessel design shows a lion and eagle attacking a fallen cow while two little bears pluck fruit from a date palm. The meaning of the imagery is unknown and may well have been interpreted very differently between the vessel's place of manufacture and its final destination.

It has been suggested that this vessel may have originated at Khafajah (the site had been looted before the arrival of archaeologists). If this is correct, at least some of its imagery may have been meaningful to the people

**FIG. 39** Chlorite vessel (roll-out of exterior) carved with mythological scenes, *c.*2600–2400 BC, diam. 17.8 cm, 1936,1217.2 © Trustees of the British Museum

**FIG. 40** Lapis lazuli cylinder seal and modern impression with human faced bison and lion-headed eagles, *c.*2500–2350 BC, ht. 4.1 cm, 1897,0511.108 © Trustees of the British Museum

of the lower Diyala as snakes were associated with the god Ninazu whose cult centre was at Eshnunna (modern Tell Asmar). Ninazu seems to have originally been connected with the realm of the dead and may even have been considered its ruler. He is later described as 'king of the snakes' and his sacred animal is a snake-dragon, the so-called *mushhushshu*. It may be significant that in Mesopotamian myth routes through the Zagros Mountains gave access to the land of the dead and indeed the Sumerian word KUR could refer to mountains, a foreign (generally eastern) land, or the netherworld.

In Mesopotamia, a mythological creature that was closely associated with the Zagros Mountains was the lion-headed eagle. This creature can be related to those known from cylinder seals and other objects of the fourth millennium BC (see fig.22). He was known to Akkadian speakers as *Anzû* which may possibly be read as Imdugud in Sumerian. The creature was represented in two postures in the third millennium BC, in a frontal view with outstretched wings, or with closed wings, its body in profile but its head lowered and viewed as if from above. It was understood as a manifestation of the thundercloud, a source of fertility but also of potential destruction. On cylinder seals, inlays, and plaques the *Anzû* is often depicted perched on the back of a human-faced bison that it bites (fig.40). This scene would seem to symbolise the rain clouds (*Anzû*) that occasionally obscured the eastern Zagros Mountains (human-faced bison). Human-headed bison generally appear on seals in pairs where they can be shown, as here, being protected by heroes and bull-men. The scale-like mountain and sprouting plant between the bison have their origins in earlier Proto-Elamite imagery (see fig.24). Indeed, it may have been the Iranian artistic tradition of depicting animals with human attributes or characteristics that stimulated Mesopotamian representations of the human-faced bison and the bull man and associated them with the east. Both creatures probably derived from the wild bison that are thought to have lived in the extensive forest cover in the Zagros ranges.

## THE ROYAL TOMBS OF UR

Relationships with the mysterious realms of the Zagros and the Iranian Plateau are found in much of the imagery and objects excavated from a group of sixteen magnificent burials at Ur dating to the period 2500–2400 BC – the so-called Royal Tombs. Ur was the centre of a powerful state dominating the alluvial plain at the head of the Persian Gulf and it is likely that these unique burials belong to the city's rulers. The vast majority of the tombs had been plundered in antiquity but the few that remained largely intact allow for some general conclusions to be drawn. Each burial consisted of a rough stone structure, some divided internally into a number of 'rooms' built at the bottom of a large pit (ranging in depth from three to ten metres). Inside the stone tombs were placed one or more bodies surrounded with finely crafted metal and stone vessels, rich jewellery, and what may be personal items. Outside the tomb, on the floor of the pit, were laid the bodies of draught animals, carts or sledges, and a large number of people who were provided with jewellery and equipment including musical instruments. In one grave alone, the so-called 'Great Death Pit', seventy-four bodies mainly of women or young girls were arranged in rows, perhaps to create a version of the ritual banquet depicted in contemporary inlays and reliefs. Recent analysis of some of the remains suggests that at least some of these individuals had been bludgeoned to death and their bodies burnt (as with sacrificial animals?) before being dressed. Perhaps they were criminals or prisoners of war – like those also known from royal imagery (see fig.37) – executed for the occasion?

Much of the jewellery in these graves is fashioned from materials that originated in the East (fig.41). Earrings were made from two pieces of sheet gold, each shaped like a hollow crescent, hammered together and fashioned over a bitumen core (now missing). They may have hung from a headdress or hair rather than being suspended from earlobes. Both the necklace and the collar (or cuff) consist of separate beads of lapis lazuli and sheet gold. The pin, made of silver with a gold and lapis lazuli cap, was used to fasten clothes and possibly to attach the owner's personal seal or amulets. The similarities among the jewellery worn by different people in the tombs suggest that much of it was manufactured at the same time, perhaps for the burial ceremony itself. Given that entrances to the netherworld were thought to be within the eastern mountains it may be that the objects in the tombs were intended to accompany the dead on that journey or be presented as gifts when they arrived there – the dead would thus be returning the stones and metals to the place where they originated in reality but now within a supernatural realm.

FIG. 41 Gold, silver and lapis lazuli jewellery from the 'Great Death Pit' in the Royal Tombs at Ur, Iraq, *c.*2500–2400 BC, pin l. 16 cm, AN1930.219–222 © Ashmolean Museum, University of Oxford

The sculpted heads of bulls ornamenting the sound box of musical instruments in some of the Royal Tombs also point to the Iranian highlands. They are provided with elaborate lapis lazuli beards suggesting that bison may be intended. In addition, examples of the human-faced bison occur, such as the pair that is represented on an inlaid plaque that decorated the front of an enormous lyre (fig.42). In the top register a nude belted hero holds two human-faced bison. The hero stands over the lower panels where the scenes depict a banquet. However, drawing on the Iranian tradition of animals acting as humans, what is presented is an 'other' worldly version of festivities, perhaps in the netherworld itself. In the lowest register appears a scorpion-man, a creature associated in later Mesopotamian literature with distant lands, mountains, and the passageways through them. As previously discussed, the human-faced bison may stand for the eastern mountains. Perhaps the hero who is dominating them is blocking or controlling access to this 'other' world. He may be comparable with the figure of the gatekeeper, known from a mythological text described below, who confronts the goddess Inana when she attempts to visit the underworld.

## FROM CITY-STATE TO EMPIRE

Although culturally unified, the southern alluvium was politically fractured with rival city-states jostling for access to resources and authority. This situation was maintained into the twenty-fourth century BC when temporary alliances or the emergence of more permanent unity, often the result of warfare, produced ever larger kingdoms. This process came to an end when Lugalzagesi, king of Umma, managed to capture Lagash (the two states had been at war on and off for over a century). If Ur and Uruk were not already part of his kingdom, he soon claimed them and presented himself in inscriptions as chosen by the god Enlil – head of the Sumerian pantheon – as ruler of 'the land', that is southern Mesopotamia. He was, however, confronted by the most powerful ruler in the northern half of the alluvium, Sargon of the city of Akkad (or Agade).

The defeat of Lugalzagesi by Sargon (2288–2235 BC) may have been viewed by their contemporaries as simply a shift of political power to yet another strong leader. Yet Sargon established a dynasty that would rule not only the ancient centres of the alluvium for a century and a half but, more significantly, would extend its control and influence well beyond the southern plains to create an empire. Future rulers would consider these achievements as an ideal to be emulated and their scribes composed literary works that wove the Akkad rulers into myths, legends, and folktales where they were either glorified or vilified. Understandably caution needs to be exercised in using such compositions to reconstruct history. Nonetheless, rare royal inscriptions of Sargon and his successors survive, as well as occasional administrative records that help to illuminate the period. In addition, the kings commissioned some of the most magnificent and complex works of Mesopotamian art, some of which make reference to the highlands.

Sargon consolidated his hold over southern Mesopotamia by building on existing political, administrative, and religious structures to forge a sense of unity between the city-states. Continuity with the past was maintained by retaining many of the defeated city leaders as local governors. Stress was placed on the belief that the individual city-state gods were all related to one another as in a large household, while Sargon and his successors gave offerings in many of the great temples but especially the Ekur ('mountain house'), home of the great god Enlil at Nippur. In addition Sargon's daughter Enheduanna was installed in the position of priestess in the temple of the moon god at Ur.

Sargon's capital Akkad has not been located but almost certainly lay on the Tigris close to where it is joined by the River Diyala. The location of the city ensured its rulers maintained an active interest in the potential

**FIG. 42** Inlay of the 'Great Lyre' from the 'King's Tomb' in the Royal Tombs of Ur, Iraq, *c.*2500–2400 BC, ht. 33 cm, B17694A © University of Pennsylvania Museum object no. B17694A, image no. 150082

TOP:

**FIG. 43** Indus Valley steatite stamp seal with bull and inscription, from Kish, Iraq, *c.*2100–1900 BC, ht. 2.4 cm, EA1931.119 © Ashmolean Museum, University of Oxford

BOTTOM:

**FIG. 44** Shell cylinder seal and modern impression of the sun god Shamash, *c.*2200 BC, ht. 3.1 cm, AN1930.112 © Ashmolean Museum, University of Oxford

threat posed by the people of the western Zagros as well as access to the riches available via the Great Khorasan Road and piedmont routes. As a result, after undertaking campaigns into northern Mesopotamia, perhaps as far as Mari and Ebla, to ensure access to the trade routes and resources of that region, Sargon turned his attention to the east. He declared he had conquered as many as thirteen regions or cities including Elam, Susa, and a place called Awan (perhaps the region around Godin Tepe) from which he claimed to have collected many spoils. Authority over the Susiana plains as well as cities like Ur gave Sargon control of the lucrative Persian Gulf trade as well as links to the Indian Ocean beyond. Indeed, in one of his inscriptions he claimed that the ships of Magan (Oman and the coast of Iran on the opposite side of the Gulf), Meluhha (Indus Valley), and Dilmun (modern Bahrain) moored at the harbour of Akkad bringing to Mesopotamia not only the resources of these regions but their people and traditions. Among the most easily portable exotic objects were distinctive stamp seals, such as this example excavated at Kish which is engraved with an image of a bull, the most popular animal to feature on Indus Valley seals (fig.43). The animal is shown in profile standing before what might be an altar or perhaps a manger. Along the top edge of the seal is an inscription in the yet to be deciphered script of the Indus Valley civilisation.

Under Sargon's sons, Rimush (2233–2227 BC) and Manishtushu (2226–2212 BC), some areas of northern Mesopotamia appear to have been placed under greater direct control and alliances established with local states. In contrast, campaigns in the east were often no more than plundering raids. Rebellion against Akkad was an ever present prospect and after crushing a revolt by a significant number of cities across the south of the alluvium, Rimush turned to deal with unrest in the highlands – frustratingly the locations of the people and places mentioned in his inscriptions remain uncertain. We do, however, get a clear sense of the prodigious wealth that the Akkadian kings could accrue from these attacks.

> *Rimush, king of the universe. He vanquished Abalgamash, the king of Parahshum, in battle. [The troops of] Zahara and Elam gathered in Parahshum, and he vanquished [them] and 16,212 men fell. … Between Awan and Susa, in the 'middle river', he heaped destruction upon them around the city. He slew the cities of Elam and destroyed their walls and uprooted the foundation of Parahshum in the land of Elam. Rimush, king of the universe ruled over Elam. … When he smote Elam and Parahshum, he brought back thirty minas of gold, 3,600 minas of copper, and 360 male slaves and female slaves and presented them as a votive to Enlil.*

Manishtushu's inscriptions describe raids against kings and cities further south in the Zagros all the way to Anshan (the region around Tal-e Malyan) and a place called Shirihum. From here we are told that a naval expedition was undertaken along the Iranian coast (or perhaps across the Persian Gulf to the Oman Peninsula) which brought the Akkadian army to a place where it was confronted by a military coalition of thirty-two cities. Manishtushu unsurprisingly claimed victory and he had diorite, the favoured stone for royal sculptures, quarried in the mountains and shipped back to Akkad.

The eastern mountains continued to hold a special place in the imagination of the inhabitants of the lowlands but representations of a mysterious 'other' world guarded by human-faced bison was largely replaced with those of Shamash, god of the sun and justice. He is one of the most popular subjects of cylinder seals of this period (fig.44). The saw-toothed blade he holds as well as the rays that rise from his shoulders generally identify the deity. Wearing a headdress with the horns of a bull to show that he is a god, Shamash can be shown seated or, as here, standing with one foot placed on a mountain, emerging between two mountains assumed to correspond to the sun rising each morning over the jagged peaks of the Zagros as seen from Mesopotamia. Shamash is accompanied by gate keepers who hold open the leaves of a giant door of heaven through which the god passes at dawn and dusk.

The Akkadian Empire reached its greatest extent under Naram-Sin (2211–2175 BC). Although he too faced challenging internal rebellions and external invasions, he claimed to rule over the 'Four Quarters', that is, lands in all directions. Indeed, Naram-Sin campaigned as far north and west as the Taurus and Amanus mountains, in the Khabur plains (ancient Subartu) where a fortified 'palace' was built at Tell Brak, and through the Persian Gulf to the Oman Peninsula. In his inscriptions, Naram-Sin places much importance on the fact that he had travelled to regions where no earlier king had ventured. Such were his successes that the people of Akkad built a temple for him and he was named god of the city.

More direct control was established over Elam during the reign of Naram-Sin and there were royal building works undertaken at Susa where bricks stamped with the king's name have been found. The integration of Susa and probably other parts of Elam into what was now a sizeable empire brought Akkadian speaking officials, administrators, and military personnel to the region. Among the cuneiform tablets discovered at Susa is a treaty between Naram-Sin and an unnamed individual who may be one of the local rulers within Elam or Awan. Written in the Elamite language, the document begins by invoking the names of Elamite gods, and also includes a few Sumerian and Akkadian deities. Old Elamite is poorly understood but one line is clear: 'The enemy of Naram-Sin is my enemy, the friend of Naram-Sin is my friend'. Diplomacy as much as military aggression was clearly a key element of the empire's strategy of control.

The entire length of the Zagros, and thus the security of the piedmont route, remained of vital importance. This is demonstrated dramatically by one of the most magnificent works of art from antiquity, the Stele of Naram-Sin (fig.45). The limestone stele (surviving to a height of around two metres) was originally erected in the city of Sippar, cult centre of Shamash, and depicts a campaign in the central Zagros. Dominating the scene is Naram-Sin himself, shown much taller than the other figures, and wearing a horned helmet to express his status as a god. The fragmentary cuneiform inscription, now barely visible in a series of lines above the king's head, recounts a

**FIG. 45** Limestone victory stele of Naram-Sin, from Susa, Iran, *c.*2200 BC, ht. 200 cm, Sb4 © RMN-Grand Palais (musée du Louvre) / Franck Raux

military expedition in the region of Lullubum against a king called Satuni and presumably also the Lullubi people and perhaps their neighbours the Guti. Lullubum was probably located in southern Kurdistan, possibly around the Shahrizor plain. The two trees represented on the stele are a variety of oak native to these mountains and reference to the region of the campaign may even include the very sandstone from which the stele is carved, perhaps quarried in the Jebel Hamrin. The traditional Mesopotamian approach of dividing imagery into a number of clearly defined registers (such as in the Uruk Vase and temple plaques, see figs 17 and 24) dissolves here to create a unified composition. The Akkadian soldiers and their enemies all gaze upwards to Naram-Sin who strides across the steep, wooded terrain with apparent ease. The Lullubi wear distinctive clothing of animal hides, with their long hair tied back. Naram-Sin's right foot presses upon two naked (and therefore dead) enemies, below whom a third naked man falls. Other Lullubi are dying or pleading for their lives, one with his spear broken, perhaps as a symbol of submission. These warriors are presented as being defeated simply by the presence of Naram-Sin rather than through his actions – it is the king who holds the power over life and death as he reaches the highest mountain peak above which stars indicate the great gods.

Naram-Sin's campaigns to maintain access to the resources of the highlands did not provide long-term security to the Akkad regime. Shar-kali-sharri (2174–2150 BC) was faced by a coalition of forces from Elam and Marhashi (a region probably to the east of Elam). A major battle took place at the confluence of the Tigris and Diyala rivers, dangerously close to the capital. Diplomacy, however, continued to play an important part in maintaining relations with these regions such as when a daughter of the king of Marhashi married either Shar-kali-sharri or his son. There was at least one campaign by the Akkadian king further to the north along the piedmont against the Guti, inhabitants of the Zagros who were raiding the lowlands – where they were accused of cattle-rustling – but also arriving in increasing numbers as settlers.

Later Mesopotamian legend credits the collapse of the Akkadian Empire to an invasion by the Guti, sent by Enlil as punishment for Naram-Sin's wrongdoings which had included the looting of the temple of Enlil at Nippur. The reality was likely to have been much more complex with leaders of the eastern highland regions pressing their advantage as insurrection arose among the city-states of the southern alluvium. Susiana may have become independent later in the reign of Shar-kali-sharri and a number of states in the Zagros chipped away at the eastern borders of Akkad. Gutians – as well as probably Lullubians and Elamites – had been settling the alluvium peacefully, perhaps for generations, and had become part of the broader multi-lingual population. As Akkad's power faltered a dynasty of Gutian rulers emerged to control several city-states across the centre of the plain (Adab and Umma) presenting themselves as the heirs of Akkad that had been reduced to controlling only the region around the Tigris and Diyala river confluence. Indeed, the situation was so confused that a later scribe compiling a list of rulers for this period exclaims: 'Who was king? Who was not king?'.

With the collapse of the Akkadian Empire the underlying independent city-state structure re-emerged across southern Mesopotamia, and trade routes adjusted to the new political order. Local independent rulers presented themselves as pious temple builders. Among the best known is the dynasty of Lagash under the rule of Ur-Bau and his son Gudea (*c.*2150 BC). In one of the longest texts that survives written in Sumerian, Gudea describes the exotic materials he used to furnish the Eninnu temple of the god Ningirsu in the city of Girsu. These were 'brought', which we should probably understand as traded, from as far afield as the Amanus Mountains in Turkey. When it came to south-western Iran, however:

> *He [Gudea] struck the cities of Anshan and Elam with weapons and brought their spoil into the Eninnu for Ningirsu. On the day when Gudea, the ruler of Lagash, had [finished] building the Eninnu for Ningirsu, he made it [the spoils] a permanent donation [to the temple].*

## KUTIK-INSHUSHINAK OF AWAN, SUSA AND ELAM

Despite these apparently successful raids from Mesopotamia, the territory of Elam was being absorbed into an emerging state that would confront the rulers of the alluvium. Towards the end of the Akkadian Empire, the region of the central Zagros had come under ever greater centralised control to form the kingdom of Awan. By around 2100 BC Susa and Anshan had been incorporated into this wide territory, which was ruled by a man named in Mesopotamian documents as Puzur-Inshushinak; it is probable that he was called Kutik-Inshushinak in the Elamite language. The final element of the king's name, Inshushinak, refers to the chief god of Susa, which may suggest that this was his home city. Cuneiform inscriptions from the time of Kutik-Inshushinak found at Susa are written in either Akkadian or Elamite and a few of them are bilingual, which may reflect an attempt by the king

**FIG. 46** Limestone statue of the goddess Narundi, from Susa, Iran, *c.*2100 BC, ht. 109 cm, Sb54 and Sb6617 © RMN-Grand Palais (musée du Louvre) / Daniel Amaudet / Jean Schormans

to integrate highland and lowland traditions. Restricted to his reign are inscriptions in a new script, known today as Linear Elamite. Although some of the signs look similar to those used in the Proto-Elamite script there is no connection between the two systems, except that both remain un-deciphered. A Linear Elamite inscription is carved on the throne of this statue representing the Elamite goddess Narundi from Susa (fig.46). According to a second inscription on the throne, but this time in Akkadian cuneiform, the sculpture was dedicated by 'Puzur-Inshushinak, prince of Awan'. The statue may be a rare surviving cult image. The goddess wears a Mesopotamian-style fleece robe and a headdress with three pairs of horns, while holding a palm frond and a vessel. Her eyes and ears would originally have been inlaid, probably with precious stones and metals.

The inscriptions of Kutik-Inshushinak that can be read provide us for the first time with a limited picture of events from an Elamite perspective. He calls himself the king of Awan and also describes himself as ensi (Sumerian 'governor') of Susa and *shakkanaku* (Akkadian 'general') of Elam. The last two titles had been used by representatives of the Akkadian kings at Susa and this might suggest that the new ruler had been in their service before coming to power. In the Elamite texts Kutik-Inshushinak is called *sunkik halmeak* 'king of the land' and *halmenik Shushim* 'sovereign of Susa'. He conquered more than seventy towns in the highlands and incorporated, or at least raided, areas in the Diyala River and Hamrin Valley (including the lands of Eshnunna, Tutub, and Akkad), perhaps seeking control of the Great Khorasan Road and the highlands beyond. At its greatest extent, his conquests may have reached as far as the Hamadan Plain in the east and perhaps laid the foundations for the future state of Shimashki.

The expansion of Kutik-Inshushinak's power onto the alluvial plain and his control of trade routes at either end may have been one of the reasons for the expansionist policies of some of the city-state rulers. Utuhegal of Uruk led the way. He claimed authority over the cities in the far south of Mesopotamia and is credited with having defeated Tirigan, the last of the Gutian kings. However, it was his governor at Ur and successor, a man named Ur-Namma (2110-2093 BC), who successfully confronted Kutik-Inshushinak's forces in the region of Akkad. Whether this blow to the kingdom of Awan was fatal is uncertain but the tables appear to have been turned and over the coming decades it would be the rulers of Ur who would extend their authority over the highlands.

## THE THIRD DYNASTY OF UR

According to Mesopotamian tradition, this was the third time that the city of Ur had dominated southern Mesopotamia – hence the modern designation of 'The Third Dynasty of Ur' or 'Ur III Period'. Gradually Ur-Namma widened his influence across southern Mesopotamia and had temples and other monumental buildings erected at a number of cities including Nippur. His most famous monuments are the massive solid mudbrick stepped platforms or ziggurats with shrines at Ur, Uruk, and Eridu. It was probably from this last site that this basalt pivot-stone originates (fig.47). A cuneiform inscription is carved delicately beside the depression in which a wooden doorpost once turned. It records in Sumerian – revived as a prestigious language of the state – the building of a temple (presumably the one in

**FIG. 47** Basalt door pivot stone of Ur-Namma, from Eridu, Iraq, *c.*2100 BC, l. 40 cm, AN2011.42 © Ashmolean Museum, University of Oxford

which the pivot-stone was installed) dedicated to Enki, god of fresh water and wisdom. It is possible that this object originates from Enki's main temple at Eridu. Ur-Namma is named in the inscription as responsible for ordering the temple to be built. As with the earlier kings of Akkad, the rulers of Ur preferred dark, hard stones for prestige and durability and fashioned their statues, vessels, and door sockets from them. They boasted in their inscriptions of bringing some of the most prized of these stones (basalt and diorite) from distant lands, the so-called mountains of Magan (southern Iran and the Oman Peninsula).

Under Ur-Namma's son Shulgi (2092–2045 BC) there was a far-reaching reorganisation of the state and a policy of expansion beyond the alluvium. This included an overhaul of the army, the creation of a centralised bureaucracy and a school curriculum, the standardisation of weights and measures, and the unification of calendars and taxation. The results of these complex new systems are recorded in tens of thousands of surviving cuneiform documents. The alluvial plain was divided into a number of provinces (largely formed from the earlier independent city-states) each administered by a governor. Each of the provinces contributed products, agricultural resources, and human labour as part of the centralised taxation system. In the east, however, a different system was developed. The region reaching from Susa along the piedmont zone and the western Zagros to beyond Assur was placed under a military government. Generals appointed by the king were responsible for collecting tribute and consequently great numbers of cattle, sheep, and goats were taken to the alluvium where they entered the tax system and were redistributed for food, temple offerings, or for processing into other commodities. This region's importance to Shulgi is suggested by his claim that 'I know the tongue of Elam like the tongue of Sumer'. Indeed, it is during his reign that there is the first clear evidence of Ur's control over Susa: bricks stamped with the king's name commemorate his rebuilding of the temple of Inshushinak in the city.

Some areas of Iran were drawn into the political and economic orbit of Ur by means of arranged marriages between the daughters of the kings of Ur and local rulers of kingdoms such as Marhashi, Anshan, and Bashime. At other times, however, these same areas became the target of Ur's military aggression, and vast amounts of plunder in the form of objects made from semi-precious stones, and metals were carried to southern Mesopotamia. Of all the highland kingdoms the most problematic for Shulgi and his successors was Shimashki, which may have lain to the north of Elam on the eastern ranges of the Zagros; the Shimashkians played a leading part in local uprisings.

## MYTHS AND REALITY

The kings of Ur had close cultic, political, and probably familial connections with the neighbouring city of Uruk as revealed in Sumerian literary texts of the period. These are focused on the earlier heroic kings of Uruk and its patron deity Inana but also reflect on-going concerns with the eastern territories. Although their meaning is not entirely clear, the texts imagine the highlands in a variety of ways. A story known today as 'Enmerkar and the Land of Aratta' has been pieced together from a number of fragmentary tablets (fig.48). It describes relations between the rulers of Uruk and a place called Aratta, somewhere on the Iranian Plateau. It is likely that Aratta does not refer to a specific place – this is, after all, a literary composition – but draws on the understanding of the real but distant, resource-rich lands. The tale may be summarised as follows:

Enmerkar, the king of Uruk, sends a messenger to the lord of Aratta to demand his submission and have skilled workers brought to Uruk – together

**FIG. 48** Fragmentary clay cuneiform tablet with the story of *Enmerkar and the Land of Aratta*, *c.*1900–1600 BC, ht. 6.2 cm, AN1924.475 © Ashmolean Museum, University of Oxford

with their precious metals and stones – in order to build shrines. The messenger travels over seven mountains (a common description to describe an immensely long distance) but when he delivers his message the lord of Aratta refuses to submit and challenges Enmerkar to a contest of wits. After several challenges back and forth, Enmerkar prepares to send a messenger again to demand that Aratta submit to him. However, the message is too long for the messenger to learn and so Enmerkar inscribes it on a tablet – writing is invented. The lord of Aratta, unable to read the inscribed signs, realises that he has been defeated by a superior intellect.

Another story that probably also reached its classic form at this time involves a king of Uruk called Bilgames (an early form of Gilgamesh) and a journey to the east (fig.49).

Bilgames seeks glory through an expedition to the Cedar Mountain, far to the east where the sun rises. Utu (the Sumerian version of the sun god Shamash), offers him seven constellations as guides. A band of young men led by Bilgames set out from Uruk and cross seven mountains until they eventually find a suitable tree to fell. On cutting down the cedar, its guardian Huwawa awakes and attacks both Bilgames and his servant Enkidu. The cunning Bilgames promises his sisters as wives for Huwawa along with other gifts. Huwawa surrenders his supernatural defences but Bilgames takes him prisoner. Complaining to Utu about this treachery, Huwawa seeks the mercy of Bilgames. However Enkidu warns of the dangers and cuts Huwawa's throat. The severed head of Huwawa is presented to Enlil but the god is furious that the guardian of the cedars was not treated with the honour that he deserved.

While the Huwawa story illustrates the importance of the eastern mountains as a source of good timber for Mesopotamia, the Zagros trade routes were increasingly providing the lowlands with the animal that would transform the history of the entire Near East and beyond: the horse. Towards the end of the third millennium BC cuneiform texts make reference to horses as the 'donkey of the mountain' revealing something of their origin. They had perhaps been introduced into Mesopotamia in connection with the trade in metals since the majority of ancient copper and tin mines found in the Eurasian steppe and in Central Asia were situated in those areas where horses had been domesticated. In the period 2100–2000 BC the bones of horses appear at Godin Tepe and Tal-e Malyan. Analysis of the teeth of both mules and horses from the latter site suggest that the animals were being controlled using a hard, probably metal, bit – a new technique quite distinct from the lip and nose rings used in Mesopotamia. In the early second millennium BC horse-drawn chariots were also introduced to Iran and Mesopotamia from Central Asia and the demand for horses soared in response to this innovation.

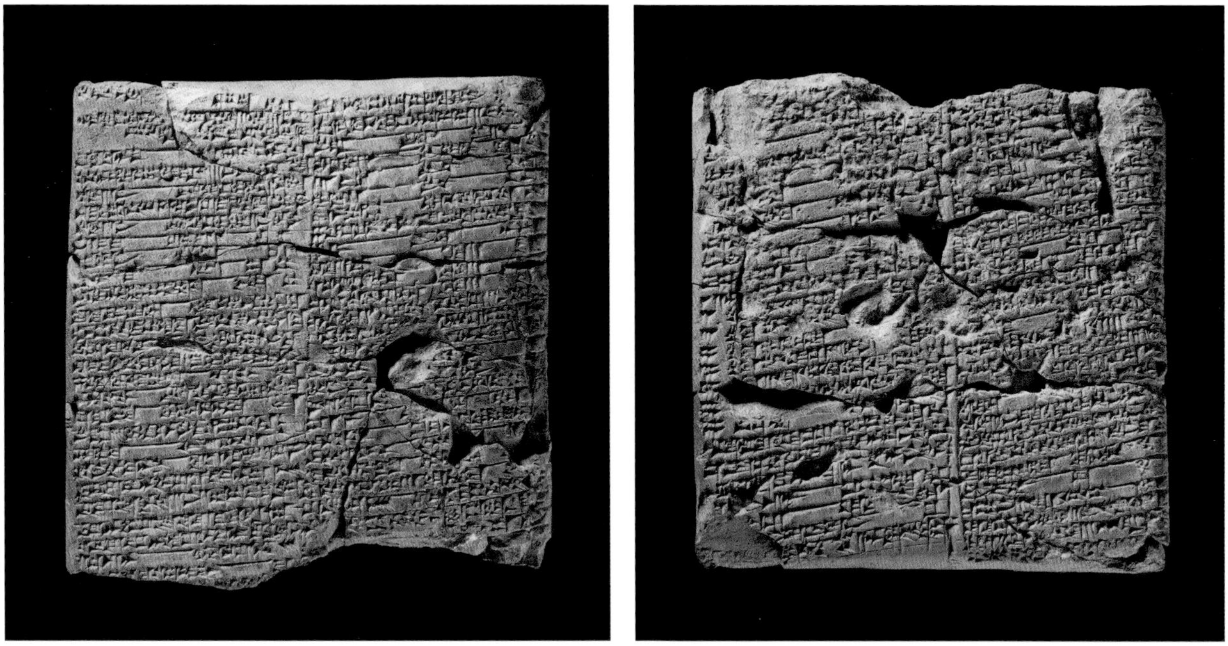

**FIG. 49** Clay cuneiform tablet with the story of *Bilgames and Huwawa*, from Kish, Iraq, *c.*1900–1600 BC, ht. 12.5 cm, AN1932.155 © Ashmolean Museum, University of Oxford

## COLLAPSE AND CONTINUITY

The reign of Ibbi-Sin (2026–2003 BC) witnessed the rapid collapse of the Empire of Ur. Centralised control was weakened by a number of factors that, according to correspondence between the king and his officials, included shortages of grain, and disruption to the routes of communication and supplies by mobile groups of pastoralists (described as 'Amorites'). This was a threat to the centralised tax and tribute system that could not be ignored. The governor of Isin, Ishbi-Erra (2019–1987 BC), took the initiative and began to extend his authority and independence.

Similarly, some of the regions under military government along the piedmont also broke with Ur. The first state to do so was Eshnunna under its governor Ituriya. A temple built for the worship of the deified Shu-Sin was incorporated into a new palace of the rulers of Eshnunna who took the title 'governor of Tishpak' – a version of the weather god Teshup, head of the Hurrian pantheon. By the early second millennium BC, Hurrian speakers (a language unrelated linguistically to others in the wider region) were found among the populations of the northern steppe lands at Nuzi, Ekallatum, Qatara, and Chagar Bazar. They had settled throughout the mountains and plains from south-eastern Anatolia across northern Syria and Mesopotamia into north-western Iran during the third millennium BC and may be connected with the Early Trans-Caucasian culture. Now rulers with Hurrian names controlled a range of small states in the Zagros such as Simurru, Tukrish, and Shusharra. A number of impressive rock reliefs in this region – the mountainous fringes of the Shahrizor Plain and the upper reaches of the Diyala River – probably date to this time. Around ten examples are known and all depict variations on the theme of the Victory Stele of Naram-Sin (fig.50; see fig.45). The majority are probably to be connected with the state of Simurru (the region of the Shahrizor Plain) and Lullubum (the highlands to the south-west of the Shahrizor). Four rock reliefs at Sar-e Pol-e Zahab may all depict Anubanini, king of the Lullubi who is named in the accompanying Akkadian inscriptions (fig.51).

Before the end of Ibbi-Sin's reign Susa had probably already been occupied by the forces of Shimashki. Descriptions in Mesopotamian documents reveal the growing threat: an official name given by state administrators to one year was, 'The stupid monkey in the foreign land struck against Ibbi-Sin, the king of Ur' (monkeys had been associated with the mountains since at least the third millennium BC). Susiana had thus been unified with the highlands of Shimashki to establish an Elamite kingdom with a dual political and territorial structure. Around 2003 BC an

**FIG. 50** Rock relief at Darband-i Gawr, Iraq, *c.*2000–1900 BC, ht. 3 m © Osama Shukir Muhammed Amin FRCP(Glasg) (Own work) [CC BY-SA 4.0 (http://creativecommons.org/licenses/by-sa/4.0)], via Wikimedia Commons

**FIG. 51** Drawing of the rock relief of Anubanini at Sar-e Pol-e Zahab, Iran. After Louis vanden Berghe *Reliefs rupestres de l'Iran ancien* (Brussels, 1983)

Elamite–Shimashkian coalition under a certain Kindattu attacked Ur – the city was occupied and Eridu, Ur, Kesh, Uruk, and Nippur were plundered. It would be a good ten years before Ishbi-Erra of Isin drove out the Elamite garrison, while his son Shu-ilishu is said to have recovered, perhaps through diplomacy, the statue of the moon god Nanna which had been removed by the Elamites and taken to Anshan. The rulers of Isin now presented themselves as the legitimate heirs of Ur (although they would never rule more than the alluvial heartland of the earlier empire). During their reign a number of literary texts were composed that describe the appalling catastrophe that had befallen Ur as well as the restoration of the city by Ishbi-Erra. The so-called 'Lamentation over the Destruction of Sumer and Ur' highlights the fate of Ibbi-Sin after the gods had abandoned the capital so as to:

> *make the people not live in their [own] dwellings, [but rather] to give them over to an evil place [and] make the Shimashkians and Elamites – foreigners – take residence [in] their dwellings, to make the enemy seize the shepherd in his own palace, to take Ibbi-Sin to the land of Elam in fetters...*

Towards the end of the century Isin's authority was challenged by the state of Larsa, about 100 kilometres to the south. The rulers of Larsa belonged to an Amorite dynasty that had served as governors for the kings of Isin. Now Gungunum (1932–1906 BC) established a rival powerbase and campaigned against Elam, claiming to have destroyed Anshan in his fifth year and a number of years later he may have taken Susa. In the eighth year of his reign Gungunum captured Ur and gained control of the valuable Persian Gulf trade. These aggressive events may have broken a period of relative peace between the alluvium states and Elam, encouraging the Elamites to reunite under a ruler with the title of *sukkalmah*, 'grand minister' (which had previously been used by the senior official at Lagash responsible for Susiana under the kings of Ur). There followed a succession of *sukkalmahs* who established themselves as prominent players in the politics of the Near East. They all had Elamite names, such as Kuk-Kirmash and Kuk-Nassur, and used a range of titles in their inscriptions including 'shepherd of Inshushinak', '*sukkal* of Elam', '*sukkal* of Susa', and the much longer '*sukkal* of Elam, Shimashki, and Susa'.

## NETWORKS OF TRADE

While Isin and Larsa vied for supremacy in southern Mesopotamia, extensive trading systems were being established between states across the wider region. The chance discovery of cuneiform tablets in a merchant's colony at the site of Kultepe (ancient Kanesh) in eastern Anatolia has illumined one such trade network that was being organised out of Assur, ruled by an independent line of kings. This was only one aspect of the complex commercial structures in operation within other parts of the Near East. The most important resource being transported across the region was tin, an essential metal for alloying with copper to produce bronze. Although the source of the tin remains unknown it was almost certainly brought to Mesopotamia from the Iranian Plateau. This is suggested by the contents of cuneiform documents found at Tell Shemshara in northern Mesopotamia, close to a strategic pass leading into the Zagros. This was the site of ancient Shusharra, capital of a small state called 'land of the gatekeeper'. Much of the tin seems not to have passed directly from the highlands at this point but had been transported along the piedmont route all the way from Susa. The Elamites probably brought tin from farther east, perhaps Afghanistan, and their control of the trade gave them a great deal of influence. It is perhaps no coincidence that during the nineteenth century BC the city of Anshan (Tal-e Malyan) grew in size to an impressive 130 hectares of settled area. To this period may also be dated two Elamite rock reliefs at Kurangun on a route from Anshan to Susa. In the best preserved carving, a male god, wearing a horned crown, sits on a coiled serpent throne as known from contemporary Elamite seals (fig.52). Behind him is a seated female also wearing a horned crown. Both deities hold a pair of serpents in their left hands. The male god raises a vase in his right hand from which streams of water flow towards worshippers.

Wealthy and powerful, Elam now allied itself with various states of southern Mesopotamia as politics demanded. The alluvium was increasingly divided between four kingdoms. Isin, its territories reduced to the centre of the plains, was sandwiched between the allied Amorite states of Babylon in the north and Uruk to the south, with Larsa dominating the extreme south at the head of the Persian Gulf. Under Kings Abisare (1905–1895 BC) and Sumu-el (1894–1866 BC), Larsa extended its authority northwards along the river systems on the east of the alluvium and weakened Isin's position still further by diverting water from canals towards its own lands.

Although its political and economic power was being strangled, Isin retained considerable prestige as the inheritor of Ur's legacy. Royal inscriptions and literary works continued to stress the state's connections with a deep historical past including updated versions of a composition that probably has its origins under the kings of Ur, the so-called Sumerian King List (fig.53). Although a number of examples of this text survive, the clay 'prism' illustrated here is the most extensive as well as the most complete version. Each of the four sides is inscribed with two columns of cuneiform script and lists in Sumerian a succession of cities in southern Mesopotamia and its neighbouring regions, their rulers, and the length of their reigns. The King List is not history as we would understand it but a combination of myth, legend, and historical information. It starts with the mythological origins of kingship when figures such as Dumuzi (otherwise known in

FIG. 52 Rock relief at Kurangun, Iran, c.1900–1600 BC, w. 4 m © UCL, Institute of Archaeology

Mesopotamian texts as a shepherd god and husband of the goddess Inana) are represented as reigning for fantastically long periods. This is followed by one of the document's most famous lines: 'The Flood swept over [the land]. After the Flood had swept over [the land] and kingship had descended from heaven [for a second time], Kish became the seat of kingship'. The List continues with the reigns of legendary rulers such as Gilgamesh of Uruk. As the King List reaches historical rulers whose reigns are attested by their royal inscriptions, the length of each reign becomes more realistic. We know from other sources that some of these rulers were actually contemporaneous and so the account is an artificial construct. The List ends with King Sin-magir of Isin (1827–1817 BC). The Sumerian King List was composed to transmit a political vision that the dominion of Mesopotamia was determined by the gods and could only be exercised by one city at a given time and for a limited period. In this way the rulers of the Isin dynasty tied themselves to powerful kingdoms of the past but also acknowledged that their authority would eventually end.

In the final decades of the century centralising forces resulted in ever larger kingdoms across southern Mesopotamia combined with a growing

1923.444

influence of Elam in their affairs. In 1835 BC Larsa came under the control of a certain Kutur-mapuk, who is described in documents as the sheikh of Yamutbal – a tribal region east of the Tigris – as well as 'father of the Amorite land'. Despite these titles referencing the Amorites, the names of Kutur-mapuk's father and daughter were Elamite and it is possible that he was acting on behalf of the *sukkalmah* in the political manoeuvrings of the alluvium. His appearance on the political scene highlights the complex relations that existed between tribal groupings and the urban-based communities and central governments. While Kutur-mapak was based at Mashkan-shapir in the east of Mesopotamia, he placed his son Warad-Sin (1824–1823 BC) on the throne of Larsa. When Warad-Sin died, his brother Rim-Sin (1822–1763 BC) replaced him and eventually inherited Yamutbal which was united with Larsa. The threat posed to the other states of the alluvium by this powerful kingdom is clear from the formation of a military coalition consisting of Uruk, Isin, and Babylon that confronted Rim-Sin in the thirteenth year of his reign. Nevertheless, this formidable force was defeated and four years later in 1805 BC Rim-Sin destroyed Uruk.

In the same period Eshnunna had expanded to control the Diyala valley as far as its confluence with the Tigris and thus became the dominant power in the central piedmont, guaranteeing its control over the rich trade with the Iranian Plateau along the Great Khorasan Road. Among the cities it conquered was Shaduppum (modern Tell Abu Harmal, a suburb of Baghdad) where Akkadian texts include a list of legal judgements (often erroneously described as a law-code) in which the ruler of Eshnunna presents himself as the protector of the weak, the dispenser of justice and righter of wrongs. Eshnunna's influence extended further under its king Naram-Sin whose authority reached as far north as the trading centre of Assur and the small state of Ekallatum. This expansion was eventually challenged by a certain Shamshi-Adad (1807–1775 BC) who, from a base in the northern alluvium, captured Ekallatum and, according to the later Assyrian King List, after three years 'removed Erishum, son of Naram-Sin from the throne [of Assur]. He seized the throne. He ruled as king for 33 years'. After taking Assur in 1807 BC, Shamshi-Adad extended his influence westwards and annexed kingdoms in the Khabur Triangle, establishing his royal seat at Shubat-Enlil (modern Tell Leilan) while his son Ishme-Dagan ruled in the east from Ekallatum. At some point Shamshi-Adad also seized Mari on the Euphrates where he established another of his sons as ruler.

Later in his reign, around 1782 BC, Shamshi-Adad moved against the Guti of the Zagros who were probably disrupting trade and communication routes. Meanwhile, further south, Dardusha the ruler of Eshnunna crossed

**FIG. 53** Baked clay cuneiform prism with a version of the Sumerian King List, *c.*1810 BC, ht. 20 cm, AN1923.444 © Ashmolean Museum, University of Oxford the Tigris in the opposite direction and invaded the region of Suhu in the Middle Euphrates. The two kingdoms maintained close relations as shown by an inscribed stele from Eshnunna that describes how Dardusha conquered the city of Qabara, located east of the Tigris between the Greater and Lesser Zab rivers. The ruler of Qabara is given the title of 'king of Arbil' (modern Erbil) which was probably a cultic centre of consequence within his kingdom. The inscription indicates that the campaign was in fact a joint enterprise by the kings of Ekallatum and Eshnunna who were united by treaty. A second, but unprovenanced, stele records the same campaign and may have been set up by Shamshi-Adad (fig.54). On one side it depicts in relief a victorious warrior, killing a falling victim with a battle axe while crushing him underfoot. This may be a scene of execution as the warrior probably held in his raised left hand the end of a rope attached by a ring to his enemy's nose. Such imagery is derived from scenes known from the time of the Akkadian Empire in the third millennium BC.

Pressure from the Guti, which was possibly threatening the movement of materials, especially tin, along the piedmont route, brought the Elamites into central and northern Mesopotamia. In 1779 BC the presence of an army of 12,000 Elamite troops belonging to the *sukkalmah* Shiruk-tuh joined with forces from Assyria and Eshnunna against the Guti. The status of the *sukkalmah* as the most powerful ruler of the period is made apparent by the mention in a letter of an official in the service of Rim-Sin to the 'great king of Elam'.

On the death of Shamshi-Adad in 1775 BC his extensive kingdom disappeared relatively rapidly. Although his son and successor Ishme-Dagan claimed that he held 'the halter of the king Elam and the king of Eshnunna', he was swept aside as the ruler of Eshnunna, Ibal-pi-el (1779–1766 BC), absorbed the eastern territories into his state. Further west the Amorite ruling house at Mari was restored under Zimri-Lim (1775–1762 BC): an extensive archive of cuneiform records from his palace provides a wide-ranging insight into the politics, economics, and social structures of the period. In one letter, for example, an official succinctly singles out the significant individuals whose authority rested on carefully constructed alliances with smaller kingdoms:

> *There is no king who is strong by himself: ten to fifteen kings follow Hammurabi of Babylon; so, too, Rim-Sin of Larsa; so, too, Ibal-pi-el of Eshnunna; so, too, Amut-pi-el of Qatna [western Syria]; [and] twenty kings are following Yarim-Lim of Yamhad [north-western Syria].*

Isin is not included in this list as Rim-Sin had already conquered the city (in 1793 BC) establishing himself as ruler of the entire lower half of the alluvium. It was considered such an important achievement that he named the remaining thirty-one years in his remarkably long reign after the event. This left only the kingdom of Babylon under Hammurabi (1792–1750 BC) as his most powerful neighbour in southern Mesopotamia.

FIG. 54 Fragment of a stone victory stele, *c.*1780 BC, ht. 49 cm, AO2776 © RMN-Grand Palais (musée du Louvre) / Jérôme Galland

## BABYLON AND THE GREAT KING OF ELAM

There now began a decade in which the rulers formed short lived unions among themselves to deal with states considered a threat to their own ambitions. Around 1769 BC Eshnunna sought an alliance with Mari that gave it control over trade routes reaching from the western Zagros across central Mesopotamia to the Euphrates River. This may have been too much for the *sukkalmah* (possibly Siwe-palar-huppak) who allied with Babylon, Mari, and probably Larsa, and attacked and plundered Eshnunna. With Eshnunna significantly weakened, Mari was able to obtain tin at a good price directly from Elam and became an important supplier of the metal to the kingdoms of western Syria and Palestine. In the extensive cuneiform records preserved at Mari are many documents charting these crucial trading connections, for example:

> *Silver and gold vessels and wine sent to Siwe-palar-huppak of Anshan and Kudu-zulush of Susa.*
> *Nine minas, twenty five shekels of tin and two minas of gold brought to Mari by Iddiyatum, chief of the Mari merchants.*
> *Ishhi-Dagan with one mina of gold sent to buy tin in Elam.*
> *Fourteen talents of tin sent by boat to the West.*

Elam was now playing a direct role in the lucrative commercial networks of northern Mesopotamia and beyond. Such was its standing that in royal correspondence the Syrian and Mesopotamian rulers described themselves as 'sons' of the 'great king of Elam' although they referred to each other only as 'brothers'. Even the powerful prince of Qatna in western Syria proposed surrendering territory to the Elamite king if the latter would fight on his behalf against a neighbouring kingdom. Evidence from Susa shows that at this time Elamite influence also reached through the Persian Gulf and across the Iranian Plateau. Enzag, the supreme god of Dilmun (Bahrain and eastern Saudi Arabia) was worshipped at Susa and stamp seals and softstone vessels (and their contents) from Magan (the Oman Peninsula) were brought to the city. Other forms of softstone containers arrived from the region of Bactria in Afghanistan and may have been by-products of the tin trade.

In 1765 BC a Babylonian army occupied two towns (Mankisum and Upi) along the frontier with Eshnunna. This was a region claimed by Siwe-palar-huppak and he ordered Hammurabi to withdraw or face an Elamite invasion. At the same time it seems that the *sukkalmah* demanded that Zimri-Lim terminate his connections with Babylon. The result was an anti-Elamite alliance between Mari, Babylon, and the kingdom of Yamhad. The oath sworn on the occasion by Hammurabi is preserved in a letter from Mari:

> *From this day forward, so long as I live, I will be at war with Siwe-palar-huppak. I will not have my servants take to the road as messengers with his servants, and will not dispatch them to him. I will not make peace with Siwe-palar-huppak separately from Zimri-Lim, king of Mari.*

The Elamite king launched a successful siege of Upi and then withdrew but returned the following year with a force of 30,000 men (from Elam, Subartu, Gutium, Eshnunna, and Malgium) and marched against the region of Sippar. Nonetheless, the *sukkalmah* was decisively defeated by the Syria–Babylon alliance and surviving sources charting Elamite involvement in Mesopotamia come to an end. Taking full advantage of the collapse of strong opposition, Hammurabi skilfully established alliances with the remaining states to subdue them one by one. In 1763 BC, with support from Mari and Eshnunna, he conquered Larsa, bringing to an end the long reign of Rim-Sin. The following year Eshnunna fell to Hammurabi (he would destroy it completely in 1756 BC by diverting water channels to flood the city). This left Zimri-Lim, his ally for the past four years, and in 1761 BC Mari also fell to Hammurabi's forces (the city walls would be demolished a few years later marking the end of the city's political significance). Hammurabi now controlled both the Euphrates and Tigris trade routes including access to the Great Khorasan Road and the resources of the east – with the shift of political and economic power to Babylon, many towns of the lower Diyala shrank in size.

## STORIES OF THE HIGHLANDS

The unification of southern and central Mesopotamia under Hammurabi ushered in a period of relative peace. Despite the state-level conflicts, the scholars of the alluvium cities had maintained their scribal traditions in the temples and schools. Many of the literary works that have survived from this period are copies made by school boys of older compositions dating to the third millennium BC, although there were also new works created for the school curriculum. While no longer spoken, Sumerian continued to be written as a prestigious literary language but Akkadian was now also used to record myths and legends. A few of these surviving stories highlight the fact that the Zagros Mountains continued to fascinate the Mesopotamian elite.

One of the most obvious references to the highlands is a Sumerian story called *Lugale*, 'O king!' that tells how the god Ninurta defeated a mountain demon and his stone army. In the following excerpt, Ninurta describes the blessings brought by the mountain after his victory and assigns roles to the conquered stones (fig.55).

> *Henceforth people shall speak of Ninhursag [lady of the mountain]. Let its meadows produce herbs for you. Let its slopes produce honey and wine for you. Let its hillsides grow cedars, cypress, juniper, and box for you. Let it make abundant for you ripe fruits, as a garden. Let the mountain supply you richly with divine perfumes. Let it mine gold and silver for you ... Let it smelt copper and tin for you, make its tribute for you. Let the mountains make wild animals teem for you.*
>
> *My king [Ninurta] stood before the esi [diorite] stone: 'They shall extract you from the highland countries. They shall bring you from the land of Magan. You shall shape strong copper like leather. When a king who is establishing his renown for perpetuity has had its statues sculpted for all time, you shall be placed in the place of libations.'*
>
> *The hero turned to the kagena [haematite] stone: 'The craftsman, expert in everything, shall value you as if gold.'*
>
> *My king turned to the dushia stone. He addressed the nir, the gug [carnelian] and the zagin [lapis lazuli]: 'You shall all be decorated with precious metal.'*

Another myth, this time an Akkadian composition, imagines the mountains as home to the Anzu bird. No longer portrayed as the benevolent lion-headed eagle of the third millennium BC, Anzu is depicted as an ugly and dangerous creature who steals the gods' Tablet of Destinies on which they write the fate of the world. Anzu flees to his home in the mountains and uses the Tablet of Destinies' powers to reverse time, transforming Ninurta's arrows back into their constituent parts. Eventually Ninurta defeats Anzu through trickery and returns the Tablet of Destinies to the gods.

Perhaps the most famous myth from this period which alludes to the eastern mountains is *Inana's Descent to the Netherworld*. For reasons that are unclear Inana decides to visit the underworld (perhaps to claim the throne). She heads east (the traditional direction taken by the dead) and, upon reaching the first of the gates to the underworld, demands that the gatekeeper allows her to enter (see fig.42):

**FIG. 55** Fragmentary clay cuneiform tablet with the story *Lugale*, from Kish, Iraq, *c.*1900–1600 BC, ht. 7 cm, AN1924.560 © Ashmolean Museum, University of Oxford

*When Inana arrived at the palace Ganzer, she pushed aggressively on the door of the underworld. She shouted aggressively at the gate of the underworld: 'Open up, doorman, open up. Open up, Neti, open up. I am all alone and I want to come in.' Neti, the chief doorman of the underworld, answered holy Inana: 'Who are you?' 'I am Inana going to the east.' 'If you are Inana going to the east, why have you travelled to the land of no return? How did you set your heart on the road whose traveller never returns?'*

As she passes through each of seven gates Inana has to remove a symbol of her power until she is finally brought naked and defenceless before Erishkigal, queen of the underworld, and turned into a corpse. Although Enki, god of wisdom, carries out a plan to restore her to life Inana is required to provide a substitute and selects her husband the shepherd Dumuzi. He is required to spend half of each year in the underworld causing the fertility of Mesopotamia's agriculture to become seasonal.

## THE COLLAPSE OF KINGDOMS

Unlike Shamshi-Adad's kingdom in northern Mesopotamia, Hammurabi's realm did not collapse rapidly on his death. Nevertheless, his son Samsuiluna (1749–1712 BC) lost control of the area of Mari in the north-west and, perhaps more ominously, there were major rebellions in the south, including a revolt in Larsa, so that by the king's thirtieth year on the throne Nippur had been lost. Gradually the southern half of the alluvium slipped out of strong centralised control and many cities may have shrunk in size or were abandoned as the lack of maintenance of the canals and irrigation channels led to an economic decline – a picture that is now being challenged by archaeological discoveries in the region. What may have been especially important for Babylon was the loss of trade connections through the Persian Gulf, particularly access to the copper of the Oman Peninsula. The broad marshes of the far south, with their rich fishing and date palm groves, came under the authority of a dynasty of 'Sealand' rulers about whom little is known. By the end of the century the city of Babylon found itself unable to mount sufficient defence to withstand a raid by a Hittite army. The Hittites had emerged in central Anatolia as a powerful state during the seventeenth century BC and, under King Mursili, had swept through the Taurus Mountains in 1595 BC onto the Syrian plains and sacked Halab (modern Aleppo), capital of the state of Yamhad. With no strong state to oppose him to the south, the Euphrates lay open all the way to Babylon. The Hittite expedition marched some 800 kilometres and plundered the city, bringing Hammurabi's dynasty to an end.

For the next century there was no strong centralised state in the region and there are thus scant sources, either textual or archaeological, to help reconstruct what was happening in both the lowlands and the highlands, let alone the relationships between their inhabitants. While urban centres certainly maintained many of their ancient traditions, largely preserved by scribal activity in temples, for many groups across Mesopotamia their identity and security was centred on household and tribal affiliations. Increasingly a tribal people identified as Kassite came to the fore. These were people speaking a common language (although one that has no obvious relationship with any others of the region) whose homeland may originally have been within the Zagros Mountains. During the first half of the second millennium BC, just as with the Hurrians and Amorites, the Kassites had settled across Mesopotamia where they were recorded as labourers but also as mercenaries and even as hostile attackers. Many had settled on the edges of cities and within villages while others had more mobile lifestyles. After 1600 BC it is possible that some of the more powerful Kassite leaders came to fill the political vacuum created by the collapse of the Amorite dynasty at Babylon – although the relationship between the Kassite tribes and the ruling family is far from clear. Quite how they came to the throne and, more remarkably, how they then extended their control over the alluvium, is equally poorly understood. It may have helped that they had no ancestral links to one particular city or region. A much later account (the so-called 'Babylonian Chronicle of Early Kings') reports that Ea-gamil, last king of the 'Sealand' went to Elam after which Ulam-Burash, brother of the Kassite king Kashtiliashu, conquered the 'Sealand'. This may have taken place around 1480 BC and the rich trade with the Persian Gulf as well as access to the area's extensive date palm plantations stimulated the economy and encouraged further stability under the new regime. From here the Kassites extended their control along the sea routes to include Dilmun (modern Bahrain) where a commercial settlement and fortress came under the control of a Kassite governor.

# IV. From Kingdom to Empire: 1500–500 BC

Fragment of a black stone *kudurru* of
Assur-nadin-shumi, *c.*699–694 BC, fig. 76

# IV. From Kingdom to Empire: 1500–500 BC

THE FACT THAT THE LAST RULER of the 'Sealand' went to Elam might suggest that there were political if not economic links between the two regions. In the second half of the fifteenth century BC a new dynasty may have emerged in Elam – though as with the rise of the Kassite dynasty at Babylon, our knowledge of the end of the *sukkalmahs* and the origin of the new Elamite rulers is extremely limited. In their Akkadian inscriptions they are referred to as the 'king of Susa and Anshan' while Elamite texts use the title 'king of Anshan and Susa'. One of the first Elamite kings of this period to leave evidence of significant building work is Tepti-ahar. He had sufficient wealth and workforce to have an elaborate mudbrick tomb and the massive mudbrick terraces of a temple precinct constructed at Haft Tepe about ten kilometres south-east of Susa. On his personal seal Tepti-ahar describes himself as king of Susa and Anzan (Anshan) and the servant of the gods Kirwashir and Inshushinak. Despite these titles it is unclear if he was king of a large unified state or was instead one ruler among many in a politically fragmented landscape with Haft Tepe as his royal centre. One clay tablet from the site is dated to the 'year when the king (Tepti-ahar) expelled Kadashman-KUR.GAL' which might refer to the Kassite king Kadeshman-Harbe I (about 1410 BC) and, if correct, would suggest that the Kassites had attempted to extend their authority into Susiana.

By 1400 BC the Kassites were in control of the whole of southern Mesopotamia, ruling from Babylon. They had effectively created a unified country which can be described as Babylonia (map 4, right) – they called

FIG. 56 Pale green limestone cylinder seal and modern impression with mountain god, fish-men, and double-headed eagles, *c.*1300 BC, ht. 4.9 cm, AN1949.899 © Ashmolean Museum, University of Oxford

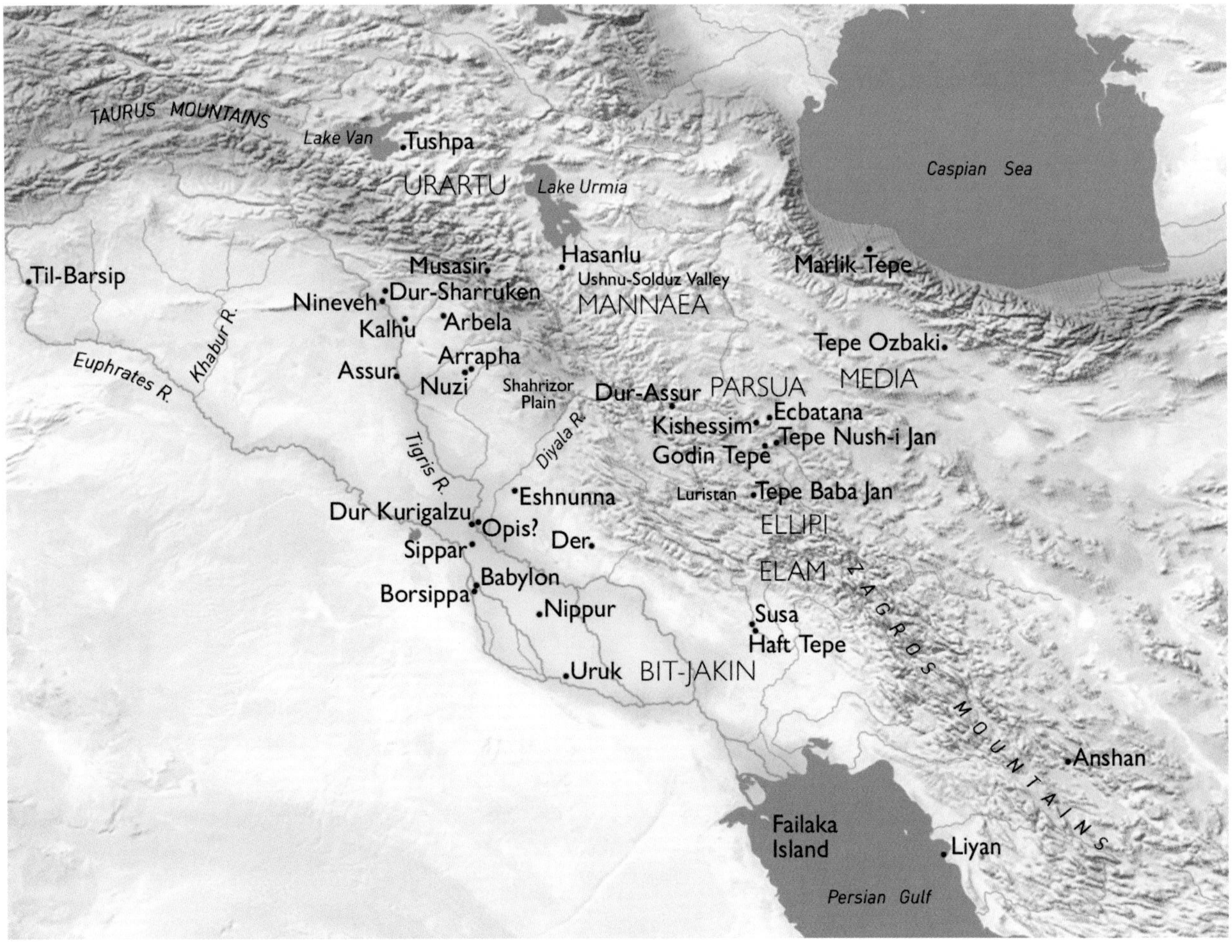


their kingdom Karduniash, which is perhaps derived from a Kassite term. As the Kassite kings extended their authority over the cities and tribal groups of the alluvium they championed, in the convention of all rulers with a claim to legitimacy, the culture of Babylonia. Indeed, ancient Sumerian and Akkadian texts were copied in the scribal schools. It is possible that Kassite identity itself was used to help establish the notion of a unique concept of kingship within southern Mesopotamia and this helped to reduce the natural tendency for the region to split into competing urban centres. The kings all used Kassite names but, although distinct Kassite gods are referenced in royal inscriptions, none of their deities were promoted and all the traditional Babylonian cults were maintained and patronised, including at Nippur where the abandoned temples were rebuilt on their old foundations. Among the earliest known examples of this Kassite piety is a small temple dedicated to Inana built at Uruk by King Kara-indash (*c.*1420 BC). The outer wall of the building was decorated with baked moulded bricks depicting bearded gods holding in clasped hands vessels from which flow streams of water. Comparable images are known from inlays and wall paintings as well as cylinder seals. This example (fig.56) carved on a pale green limestone cylinder depicts a mountain god, as indicated by the shape and pattern

FIG. 57 The partially restored ziggurat at Dur-Kurigalzu, Iraq, *c.*1400–1300 BC, ht. 57 m © Nico Tondini/Robert Harding World Imagery/Corbis

of his robe, between two fish-men. All three figures hold flowing vases. There are more vases between the figures, and their streams flow out to or from the mouths of double-headed eagles (such creatures are known much earlier from the imagery of eastern Iran). Presumably the mountains were understood as a source of water in rivers and rain. The cuneiform inscription is a prayer to Marduk, patron god of Babylon.

For the Kassite rulers the link to the mountains through the Diyala region was of key importance as it ensured them a supply of highly prized horses and lapis-lazuli as well as other resources from the highlands. To guard the alluvium end of this crucial route the royal city of Dur-Kurigalzu (modern Aqar Quf) was constructed. Its name meaning 'Fortress of Kurigalzu' suggests that the city was founded by Kurigalzu I (*c.*1390 BC), although the late fourteenth-century BC ruler Kurigalzu II also undertook work here. The site is dominated by a large ziggurat. Today this is preserved only as an exposed central core rising to a height of around 57 metres – the lowest level of which has been reconstructed – that reveals layers of reed-matting and bitumen separating every seventh or eighth course of bricks (fig.57). The ziggurat sat within a complex of courtyards and rooms similar in layout to the temple precinct at Ur, which Kurigalzu I may also have restored along with numerous other religious buildings in southern Mesopotamia. About one kilometre to the north-west of the ziggurat was a large palace. In one wing of the building (dating to late in its history) were elaborate wall paintings of rosettes as well as geometric motifs and simple bands of colour. At some doorways leading into courtyards were paintings of male figures walking in procession. Outlined in black, with black hair and shoes, the men are shown wearing white robes, with details in red, and fez-like hats.

A snapshot of life in this period is provided by the evidence excavated at Nuzi (modern Yorghan Tepe), a small town in northern Mesopotamia lying to the east of the Tigris within the kingdom of Arrapha. Texts discovered here reveal that it was populated largely by Hurrians but included Kassites, Assyrians, and other groups. The local economy was based on agriculture, animal husbandry, and trade organised by both private individuals and the state. Objects were imported both for local consumption and for re-export along the trade routes, and there was small-scale production of pottery, metalwork, and glass within the town. Indeed, Nuzi was typical of the walled settlements that dotted the landscape, and its mayor maintained relations with neighbouring settlements and people. By 1400 BC it lay in the shadow of the much larger kingdom of Mittani which, from its heartland in the Khabur Triangle, had come to dominate a geographical area reaching from north-western Syria to the plains east of the Tigris. The latter had been conquered by Saushtatar (*c.*1410 BC), who, according to later accounts, also plundered Assur and had the gold and silver doors of the city's main temple carried back to his own capital. For Mittani, apart from the control of regular trade, tribute, and taxes, one of the principal attractions of dominating such towns in northern Mesopotamia was the supply of horses which were traded through them from the east.

## DIPLOMACY AND CONQUEST

Increasingly these long-distance commercial networks linked the larger states of the wider Middle East and diplomatic relations included correspondence between the royal households. Something of these international relations during the period approximately 1360–1330 BC can be reconstructed from the so-called Amarna letters. These clay tablets, written in Akkadian cuneiform and discovered in Egypt at the site of Tell el-Amarna, represent the correspondence of Egypt's vassal rulers in the small states of the eastern Mediterranean as well as several letters sent to the Egyptian king by rulers of the other great powers including Mittani and Babylonia. Diplomatic protocol required the exchange of rich gifts and the much prized horses and lapis lazuli – described as 'lapis lazuli of the mountain', that is the genuine stone, in contrast with 'lapis lazuli of the kiln', a kind of coloured glass – travelled west to Egypt in exchange for gold. Of particular concern in the letters are royal marriages between the Egyptian king and foreign princesses with the associated bridal gifts and dowries. Thus the Pharaoh Amenhotep III (1390–1353 BC) received as brides the daughters of Shuttarna II of Mittani (*c.*1380 BC) and Kadashman-Enlil of Babylonia (*c.*1374?–1360 BC), while his successor, Amenhotep IV/Akhenaten (1353–1332 BC), married a daughter of the Kassite king Burnaburiash II (1359–1333 BC).

At some point around 1340 BC there was another state-level marriage when one of Burnaburiash's sons married the daughter of the Assyrian king Assur-uballit I (1353–1318 BC). By this time the Hurrian kingdom of Mittani had been eroded by onslaughts from the Hittites in Anatolia and their hold over northern Mesopotamia was sufficiently weakened for Assur-uballit to annex the important grain growing regions to the north and east of his capital Assur which included the towns of Nineveh and Arbela. This brought Assyrian armies to the foothills of the Zagros and the border with the state of Lullubum. It is likely that these achievements, and the potential threat they posed, were recognised in Babylon and a diplomatic royal marriage was arranged. The couple produced a son, Karahardash, who was appointed king of Babylon on the death of Burnaburiash (in 1333 BC) but he was almost immediately killed in a revolt at court. The murder of his grandson produced a swift response from Assur-uballit, who invaded Babylonia and the following year he installed Kurigalzu II (1332–1308 BC), a member of the Babylonian royal family, on the throne. The new king undertook large-scale building work at the site of Dur Kurigalzu. Kurigalzu also claims in his inscriptions to have conquered Susa, Elam, and Marhashi – he certainly seems to have refurbished temples at Susa since his brick inscriptions have been found there.

These military incursions into Elam may have helped to establish a new ruling dynasty there under Igi-halki. The dating of Igi-halki and his successors is fraught with difficulties and some scholars would place them a century earlier. However, recent work suggests they should be situated in the thirteenth century BC, although the years of their reigns remain only approximations. What is clear is that, as was the case with many of the powerful states of the wider Near East, the Babylonian and Elamite royal houses were tied by marriage. Thus Igi-halki's successor, Pahir-ishshan, married a sister of Kurigalzu II and one of Kurigalzu's daughters was wed to the Elamite king Humban-numena (about 1290–1270 BC). The close relationships between the two states may be reflected in activity in the Persian Gulf. After about 1300 BC Kassite administration was transferred

**FIG. 58** Bronze and copper statue of Queen Napir-Asu, from Susa, Iran, *c.*1270–1250 BC, ht. 129 cm, Sb2731 © RMN-Grand Palais (musée du Louvre) / Franck Raux

from Dilmun (Bahrain) to Failaka Island at the head of the Persian Gulf and it may have been as a result of negotiations with Elam over spheres of interest that Humban-numena was able to undertake building projects at Liyan (near modern Bushehr) on the opposite coast.

When Humban-numena died he was succeeded by his son Untash-Napirisha (*c.*1270–1250 BC) who married a Kassite royal bride. This king is best known for his building activities especially at Susa where he built shrines and temples for eighteen different deities. The main shrine of Inshushinak on the western edge of Susa's Acropole mound was refurbished by Untash-Napirisha and it must have been a truly magnificent building judging from the evidence for a crenellated roof and decorative polychrome inlaid terracotta wall knobs. Close by was another temple, that of Ninhursag ('lady of the mountain'), a Mesopotamian goddess. Here was discovered a masterpiece of ancient bronze casting, the 1.29-metre-high statue of Untash-Napirisha's wife, Napir-Asu (fig.58). This marvellous work of art is one of a number of large bronzes of this period discovered at Susa. The sculpture,

FIG. 59 The entrance to the vast ziggurat of Untash-Napirisha at Choga Zanbil, Iran, c.1260 BC, W. 105 m © Will Pryce/Thames & Hudson/Arcaid/Corbis

which may represent the king's Kassite bride, is formed from a copper outer shell made by the lost-wax process laid over a solid bronze core. Her short-sleeved garment is decorated with chased circles and the long skirt that wraps around her body ends in a long wavy fringe. Another fringed garment is worn over the upper half of the skirt at the back and sides. Grooves along the statue's arms and either side of the skirt may have held gold or silver foil in place. Napir-Asu's hands are held elegantly at the waist, her right hand over the left as typical of temple votive figures. An inscription in Elamite on the front of the skirt names her and curses anybody who might destroy the statue:

> *I, Napir-Asu, wife of Untash-Napirisha. He who seizes my statue, who would smash it, who would destroy its inscription, who would erase my name, may he be smitten by the curse of Napirisha, of Kiririsha, and of Inshushinak, that his name shall become extinct, that his offspring be barren, that the forces of Beltiya, the great goddess, shall sweep down on him.*

Along with his building work at Susa, Untash-Napirisha founded a completely new city called Al-Untash-Napirisha (modern Choga Zanbil) about 40 kilometres south-east of Susa. It covers an area of about 100 hectares and was enclosed by a four-kilometre-long wall. In the eastern corner an elaborate gateway led to a series of palaces. Much of the site was never built on but it was dominated by a walled area containing numerous shrines and temples which, according to associated brick inscriptions, were dedicated to some twenty-five Elamite and Mesopotamian gods. At the very heart of this sacred space, behind a further enclosing wall, was an enormous ziggurat dedicated to Napirisha and Inshushinak, the gods of Anshan and Susa (fig.59). It was built from millions of sun-dried bricks with a two-metre-thick 'skin' of baked bricks that protected it from the elements. The ziggurat was square in plan and, although now heavily eroded, may originally have consisted of five or six layers of increasingly smaller squares to create a stepped structure approximately 60 metres high (each side of the lowest level is 105 metres long). Unlike Mesopotamian ziggurats, access to the different levels was via internal staircases. A shrine on the summit was decorated with glazed blue and green bricks with touches of gold and silver. It was certainly one of the most impressive buildings ever constructed in the ancient Near East and demonstrates the expertise and size of the labour force that was available to the Elamite ruler.

## EXTENDING THE BORDERS OF ASSYRIA

The relations that linked the royal families of Elam and Mesopotamia did not prevent tensions at the borders, especially between Assyria and Babylonia. Under Adad-nirari I (1295–1264 BC) the Assyrians pushed their frontier southwards to the region of the Diyala River and thus the Great Khorasan Road, an action which can only have been viewed with alarm by the Babylonians. Meanwhile in the north and east Assyria's borders abutted mountains where the ever present threat from transhumant pastoralist groups disrupting communications and supplies was managed through a series of military campaigns. To the west, the ruler of the remnants of the kingdom of Mittani was forced to recognise the Assyrian king as his overlord through regular payments of tribute. This steppe land was brought under even tighter Assyrian control under Shalmaneser I (1263–1234 BC) with the installation of governors at a number of strategic centres. This growth of Assyria into a regional power during the thirteenth century BC is reflected in the designs of indigenous cylinder seals. Where previously seals had been

**FIG. 60** Pink limestone cylinder seal and modern impression with a winged goat, *c.*1300–1200 BC, ht. 3.8 cm, AN1938.79 © Ashmolean Museum, University of Oxford

carved with scenes of an animal pacing calmly towards a tree, they now became much more dynamic with the animal, sometimes transformed into a supernatural creature, leaping or gambolling. This pink limestone seal and modern impression is a good example, depicting a winged goat, leaping in a mountainous landscape covered with trees – perhaps derived from the crooked tree on the ancient and revered Stele of Naram-Sin (see fig.45) – and, typically, one of its hind legs is thrown high in the air with its tail outstretched (fig.60).

By the reign of Tukulti-Ninurta I (1233–1197 BC) Assyrian authority reached from the Euphrates River in the west – from which position the king's forces were able to confront the Hittites effectively – to the Zagros foothills in the east. Wide-ranging military action in the mountains to the north and east was undertaken to maintain security as well as to ensure the small states of the region directed the flow of the ever crucial tin and horses, as well as luxurious lapis lazuli, towards Assyria. This provided the tiny commercial centres of the highlands with a degree of stability, albeit at the point of a sword, as well as ensuring a market for the materials that they could supply through their eastern trade networks. In an inscription from Assur, Tukulti-Ninurta recounts his conquests in the arc of mountains reaching from the Lesser Zab (the region east of Arrapha) to the headwaters of the Euphrates:

> *At that time, from Tulsina [and] Mount Lashqu, between the city of Shasila [and] the city Mashhat-sharri on the opposite bank of the Lesser Zab, from Mount Zuqushku [and] Mount Lallar – the district of the extensive land of the Qutu [successors of the Guti] – the entire land of the Lullumu [Lullubu], the land of Paphu to the land Kadmuhu to the border of Nairi [upper Tigris] [and] the border of the land Makan, to the Euphrates – those regions the great gods allotted to me. All my enemies [I brought] under one command.*

The Assyrian king now had access to considerable resources and Tukulti-Ninurta undertook impressive building projects at Assur and founded a magnificent new city, Kar-Tukulti-Ninurta, across the Tigris from the capital. The new royal centre covered an area of around 500 hectares, with a spacious palace in one area built on a twelve-metre-high mudbrick terrace that overlooked a lower palace complex. There was also an important temple dedicated to the god Assur with an enormous ziggurat. To irrigate the fields surrounding the city, a canal was dug to bring water from the mountains. In his inscriptions the king claims that the city would serve as a treasure house for the accumulated wealth of his conquests.

As all Mesopotamian rulers of the past had discovered, direct control of the mountain regions was virtually impossible due to the sheer challenge of traversing the terrain as well as the character of the scattered communities across it. Nonetheless, Tukulti-Ninurta's control of the terminus of the trade routes in the foothills, as well as his redirection of the flow of resources from the highlands in Assyria's favour, threatened to cut supplies to Babylonia. The Kassite king Kashtiliashu IV (1232–1225 BC) reacted by leading an army north to capture Assyrian territory east of the Tigris. If the Babylonian king calculated that the Assyrians would be unable to respond because Tukulti-Ninurta was campaigning in Hittite territory he was to be proved disastrously wrong. Having consolidated his hold over south-eastern Anatolia, Tukulti-Ninurta turned to confront the Babylonian army. According to Assyrian royal inscriptions the ensuing battle was a decisive victory for Tukulti-Ninurta and the king claimed to have personally captured Kashtiliashu, who was taken in chains to Assur. The city of Babylon was plundered; strategically and commercially important areas of Babylonia were annexed to Assyria. Tukulti-Ninurta took the throne of Babylon but ruled through a succession of puppet kings.

## A RESURGENT ELAM

The Assyrian domination of Babylonia posed a challenge to the Elamites. King Kiden-Hutran (*c.*1230–1210 BC) not only faced the loss of lucrative trade routes but, since he was married to a Kassite princess, he also considered that he had a rightful claim to the Babylonian throne. This is possibly reflected in the form of a letter addressed to the Kassite court. Although the name of the author is not preserved, it may be Kiden-Hutran (the later Elamite king Shutruk-Nahhunte is another candidate). The writer of the letter claims to be the son of a Kassite royal mother and had wed a Kassite princess. He goes on to say that if his demands were not met he would 'destroy your cities, demolish your fortresses, stop up your [irrigation] ditches, cut down your orchards'. Kiden-Hutran's claims must have been rejected because around 1224 BC he advanced with an army into Babylonia, plundering the city of Der before taking Nippur and removing Tukulti-Ninurta's appointee from the throne. Nonetheless, Assyria retained its control of the region and another of their representatives was installed in Babylon. Approximately five years later Kiden-Hutran returned, more cities in Babylonia were attacked and the Assyrian appointee was again removed.

It is possible that these events in Babylonia led to a rebellion at the Assyrian court for in 1197 BC Tukulti-Ninurta I was assassinated. He was followed on the throne in fairly rapid succession by two of his sons and another claimant. While these court intrigues may not have led to any appreciable loss of Assyria's territories in Syria and northern Mesopotamia, in Babylonia the Kassite royal family was re-establishing their authority under Kashtiliashu IV's son, Adad-shuma-usur. By around 1190 BC he felt sufficiently strong to mount a challenge to Assyria's control of northern Babylonia and, having been victorious in battle, Adad-shuma-usur regained the throne for his dynasty.

Elamite attacks on Babylonia were revived under Shutruk-Nahhunte (1185–1155 BC). In about 1158 BC, and despite fierce opposition, the army of the 'king of Anshan and Susa' ransacked Babylonia and the Kassite king Marduk-apla-iddina I (1171–1159 BC) was overthrown. Shutruk-Nahhunte received vast amounts of tribute from the cities of Babylonia which he brought back to Susa. Here he undertook an ambitious building programme, judging from his inscribed bricks at the site. To the south of the Inshushinak temple on the Acropole Mound he had a building erected in which he deposited revered royal monuments that he had gathered from conquered cities in Mesopotamia, such as Sippar, Akkad, Eshnunna, and Dur-Kurigalzu, as well as some from within Elam itself, including Anshan and Al-Untash-Napirisha. These included stone records of Kassite land grants (so-called *naru* or *kudurru*) and monuments famous in their own time such as a statue of Manishtushu and the Victory Stele of Naram-Sin (see fig.45). This was more than simply the taking of war trophies but a common practice among victorious kings, since the monuments were believed to contain within their images and texts the very essence and power of the rulers who had created them, and were considered direct links to the gods who had sanctioned their creation (and had permitted the later kings to remove them). As a result a number of monuments were preserved and inscribed – on the Stele of Naram-Sin, for example, a blank space on the mountain peak was filled

**FIG. 61** Basalt stele inscribed with the law code of Hammurabi, from Susa, Iran, *c.*1792–1750 BC, ht. 225 cm, Sb8 © RMN-Grand Palais (musée du Louvre) / Franck Raux

with an Elamite inscription that names Shutruk-Nahhunte, declares that the monument was taken from Sippar, and dedicates it to the god Inshushinak – while others were recarved or defaced, reworking their power to new ends. Although not inscribed by the Elamite king, another monument that had been brought to Susa (though whether by Shutruk-Nahhunte is unclear) was the famous Law Code of Hammurabi (fig.61). It too came from Sippar, the city of Shamash, the god of the sun and of justice, who is shown in relief at the top of the stele seated on a throne, itself resting on mountains behind which the sun rose. He is passing the rod and ring, symbols of kingship, to Hammurabi, the eighteenth-century BC ruler of Babylon. The royal judgements inscribed on the stele had been considered worthy of copying within Babylonian scribal schools for centuries after the time that it was written, and so the monument came to Susa with significant scholarly as well as religious and royal prestige.

Towards the end of Shutruk-Nahhunte's reign a Kassite king was able to install himself briefly in power in Babylonia, but the Elamites returned under Shutruk-Nahhunte's son Kutir-Nahhunte (1155–1150 BC), who attacked the region and again plundered many of the major cities – some of the monuments discovered at Susa may have originated from this campaign. Later Babylonian propaganda claims that he committed the ultimate crime of removing the statue of Marduk from his temple at Babylon. These were among the final acts in a century of almost continuous conflict between Babylonia, Assyria, and Elam which fatally weakened the Kassite dynasty's long tenure of the throne. The last Kassite king, Enlil-nadin-ahi was carried off to Elam and a non-Babylonian was established as governor.

Kutir-Nahhunte was succeeded on the Elamite throne by his brother Shilhak-Inshushinak (*c.*1150–1130 BC) who describes himself in an inscription as 'enlarger of the kingdom'. Elam was now at the peak of its power and wealth as perhaps demonstrated by the growth of trading centres in the region of Anshan. Administrative cuneiform tablets from Anshan of around this time make reference to an intensive and large-scale trade in metals – gold, silver, tin and copper – and a large building at the site may have been for their redistribution and storage. The availability of ample supplies of metals may be linked to an expansion of the Elamite kings through the Zagros. Indeed, Shilhak-Inshushinak campaigned widely, particularly in eastern and north-eastern Mesopotamia where many of the territories, which had previously fallen within Assyrian and Babylonian spheres of influence, were organised into provinces. Although control over the Babylonian plains was never consolidated – it may have been considered unnecessary given there was no strong opposition – Elamite administrators were established across the eastern mountains and the piedmont to beyond the Diyala River into the Hamrin valley, allowing the Elamites to control not only the western end of the Great Khorasan Road but also the important trade routes leading to Arrapha and Assur. This posed a serious threat to Assyria under Assur-dan I (1168–1133 BC), who attempted to ward off Elamite expansion through a military campaign into the Diyala region. However, although the Assyrian king managed to deport people from the area, the Elamite border remained dangerously close to the much reduced Assyrian heartland.

Assyria was now being squeezed between Elam to the south and east and by mobile pastoralists in the west, especially Aramaean tribal groups speaking the Semitic language Aramaic. The weakening of Assyrian authority over the northern Mesopotamian steppe lands took place in the aftermath of the rapid collapse of the Hittite Empire in Anatolia and the retraction of Egyptian power in territories along the eastern Mediterranean.

Systems of control and interaction disintegrated across the region perhaps exacerbated by failing rainfall and poor harvests. Economies crumbled leading to disposed and impoverished people pressing into urban centres and the more mobile populations began to raid trade caravans and rural communities. Aramaean groups started to penetrate Assyria, attracted by the fertile lands around the Tigris Valley. The Assyrian king Assur-resh-ishi I (1132–1115 BC), who described himself as the 'avenger of Assyria', mounted repeated campaigns against these people.

Babylonia appears to have avoided any major influx of people from outside the region even though there was probably a lack of sufficiently organised forces to oppose them following the Elamite attacks. The country, however, did not break apart into the traditional rival city states, which emphasises the extraordinary success of the Kassite rulers in unifying the ancient cities and tribal lands. Their legacy was taken up by the rulers of Isin in the far south of the alluvium. The region may have avoided the worst effects of the Elamite campaigns, or perhaps the dynasty established by Marduk-kabit-ahheshu (1154–1141 BC) had been accomplished with their support. By the reign of Nebuchadnezzar I (1125–1104 BC), however, Isin felt sufficiently strong to launch attacks on both Assyrian and Elamite territory. The best known of these campaigns took place, according to a record by one of the king's officers, in the heat of summer when Nebuchadnezzar led the Babylonian army as far as Susa itself. A major battle ensued and Nebuchadnezzar's forces defeated those of the Elamite ruler Hutelutush-Inshushinak (about 1130–1110 BC). According to the Babylonian account, the Elamite king simply 'disappeared', but it is possible that he may have retreated to Anshan in the highlands while Nebuchadnezzar plundered Susa. The triumph for Babylon was celebrated in literary texts that would be copied for generations, especially because the statue of the god Marduk was said to have been recovered and returned to its temple. Nebuchadnezzar now adopted royal titles in the style of Sargon of Akkad and Hammurabi, claiming the authority of Mesopotamia's greatest rulers.

While Babylonia's borders appeared secure, Assyria was faced with a threat from Aramaean tribes advancing from the west while to the north Mushki and Kaska people were moving through the mountains, following the course of the Tigris towards the lowlands. Tiglath-pileser I (1114–1076 BC) claimed to have successfully led the Assyrian army against the northern tribesmen, taking more than 6,000 prisoners. He pursued others into the hills where he broke apart the tribal confederations and plundered their villages and livestock. While there were military successes in the mountains, the campaigns directed against the pastoralist Aramaeans on the northern plains of Iraq and Syria were less decisive in their outcome. These people now dominated the Euphrates region south of Carchemish and often raided as far east as the Tigris. The pressure on Assyria's borders was relentless and the kingdom was further threatened by Babylonian advances under Marduk-nadin-ahhe (1099–1081 BC); the city of Ekallatum, close to Assur, was captured by the Babylonian army. Tiglath-pileser was eventually able to expel the Babylonians, but only after nearly a decade, when he took his revenge by plundering Dur-Kurigalzu and Babylon. By this time, however, Babylonia was also facing Aramaean incursions and its military efforts were directed against the bands of refugees and marauders pushing south, following the Euphrates into the rich agricultural lands of the alluvium. By the time of Tiglath-pileser's death in 1076 BC Aramaean groups may even have reached the heartland of Assyria.

A lack of unity among the encroaching tribes meant it remained possible for Assyrian forces to push some of them back. Alliances were made with others or their leaders bought off. Nonetheless, the Assyrian kingdom was increasingly reduced to its homelands around Assur, Nineveh, and Arbela. As stretches of land were ravaged or claimed by tribes throughout Syria and Mesopotamia, there were notable disruptions to overland trade and links between Assyria and Babylonia and the east were severely curtailed. On the Babylonian plain there were significant attacks by substantial tribal confederations against cities such as Nippur and Sippar; the temples of the sun god Shamash were so badly devastated that the cult was not resumed for nearly two centuries. Over the decades after 1050 BC, however, climatic conditions gradually improved, encouraging the mobile pastoralists to adopt more sedentary lifestyles and the Aramaeans gathered around their leaders in marginal areas between the ancient cities. Other large tribal groupings called the Chaldaeans established themselves in the south of the country, claiming the Sealand and its control of the vast date plantations, and important trade links with the Persian Gulf.

It is difficult to follow the fortunes of the Elamites during this period as, following Nebuchadnezzar's attacks on the region, local written sources dry up. Over a century or more, a large number of towns in Susiana were deserted including both Al-Untash-Napirisha and Susa, while further south Anshan's population declined and the city was then abandoned. Their collapse was probably linked to the general crisis that had enveloped the Near East associated with the movements of large numbers of mobile pastoralists. In the mountains of south-western Iran the changes in settlement have been associated with the arrival of new peoples which may have included the ancestors of the later Persians. Indeed, it is to this period

that we should date the arrival onto the plateau of populations speaking Iranian languages. They do not become visible to us until the early ninth century BC when Assyrian administrators begin to record Iranian names among the people of the Shahrizor region. After this time, texts relating to Assyrian military activities within and beyond the Zagros reveal that the percentage of Iranian names is lowest in the west and highest in the east. This strongly suggests that Iranian tribes migrated onto the plateau via Central Asia, east of the Caspian Sea (modern Turkmenistan) although it remains possible that some groups entered through the Caucasus.

The Iranian migration probably took place in several waves and along different routes, but Iranian tribes may have been established throughout the northern and central plateau by the beginning of the first millennium BC. Nevertheless, there remained much regionalisation and local identity. The vessel illustrated here, for example, belongs to a distinctive ceramic tradition of north-western Iran known by the term Amlash, the name of the market town to which they were brought to be sold to dealers at the south-western corner of the Caspian Sea (fig.62). Dating to the late second or early first

**FIG. 62** Burnished pottery 'Amlash'-style vessel in the shape of a zebu (bull), *c.*1000 BC, ht. 39 cm, AN1964.347 © Ashmolean Museum, University of Oxford millennium BC it is characteristic of hollow containers fashioned in the shape of an animal. Here the form of a humped bull, or zebu, made of reddish brown clay has a muzzle that projects as a beak-like snout. Similar vessels have been excavated from tombs at Marlik Tepe where there were also stags, rams, and equids, along with weapons and jewellery. The containers were clearly intended as pourers but their exact function remains uncertain, although a ceremonial use is likely.

During the tenth century BC the generally unsettled conditions across Mesopotamia continued to limit the activities of the older centres of power. These problems were further exacerbated by famine. With Aramaeans dominating the Euphrates route, centralised control of much of Babylonia remained problematic. Various local dynasties claimed the throne of Babylon but throughout the century there were continual disruptions in the performance of the so-called New Year Festival which legitimised the Babylonian king – claimants were often simply unable to reach Babylon safely. According to a later text, one of these rulers, Mar-biti-apla-iddina (984–979 BC), had ancestral links with Elam but there is no evidence that he maintained any connections with the region, which remains shrouded in obscurity.

## THE REESTABLISHMENT OF ASSYRIAN AUTHORITY

The settling of the tribes gradually reduced the pressure on Assyria and King Assur-dan II (934–912 BC) launched regular military campaigns against them. Territory taken by the Aramaeans at the end of the eleventh century BC was reclaimed and, according to the royal inscriptions, people who had been driven out of Assyria through want and hunger were brought back and settled. Under his son, Adad-nirari II (911–891 BC), the Assyrian advances continued against the tribal groupings in the region of the upper Tigris and Greater Zab River, but campaigns now ranged further reaching the border of Lullubum, as well as south along the piedmont to safeguard the mountain routes. Although the Assyrians did not enter the Babylonian plain, there were clashes with the Babylonian army which was asserting its rights over this crucial region. Babylonia had also begun to recover slowly from the earlier disastrous conditions. The revival led to a reopening of diplomatic relations between Assur and Babylon and, late in his reign, Adad-nirari exchanged daughters in marriage with the Babylonian king Nabu-shuma-ukin (*c.*895 BC).

Assyrian attempts to ensure their security against the inhabitants of the neighbouring highlands – as well as access to the horses and metals of the region – were maintained under Tukulti-Ninurta II (890–884 BC). East of the

FIG. 63 The Greater Zab river with mountains in the distance, photograph taken in 1951, 2004.130.32149.1 © Pitt Rivers Museum, University of Oxford

Tigris he pushed Aramaean tribes south of the Lesser Zab River which now acted as the southern border of Assyria. The king also maintained a strong Assyrian presence in eastern Syria and was able to march through the region without apparent opposition. A sophisticated system for mobilising troops and then supplying their needs through tribute demands was established in this period. This would form the basis for much of Assyria's future military success. When Assurnasirpal II (883–859 BC) succeeded to the throne of Assur the kingdom was therefore already established as the principal power in northern Mesopotamia and eastern Syria. Like his two royal predecessors he first moved to ensure the mountainous territory to the north and north-east of Assyria remained under his authority. During the first year of his reign Assurnasirpal led his forces along the valley of the Greater Zab River, the route linking Assyria with the region south of Lake Urmia (fig.63). In 882 BC the army headed north along the Tigris into the mountainous Nairi lands, which were divided among numerous local leaders, and, following an arc westwards, the Assyrians emerged from the mountains onto the Syrian plains around the river Khabur. Here the various local rulers presented Assurnasirpal with rich gifts as expressions of goodwill.

The pretext for Assurnasirpal II's campaign in 881 BC was a rebellion by a certain Nur-Adad, the Aramaean chieftain of Dagara to the east of Arrapha. It had resulted in local rulers forming an anti-Assyrian alliance and obstructing the Bazian mountain pass (between Arrapha and the

Sulaymaniyah Plain), the entrance to the land of Zamua (with the Shahrizor Plain at its heart). According to Assurnasirpal's inscriptions his armies swept the opposition aside, plundering their walled towns and taking enormous amounts of booty. In these inscriptions Iranian names are recorded for the first time among the inhabitants of the region. The king pushed on to the headwaters of the Diyala River where Dur-Assur (possibly the modern site of Bakrawa) was established as a strategic base in the central Zagros Mountains:

> *At that time in the land Zamua, the city Atlila which Sibir king of Karduniash [Babylonia] had captured, was dilapidated [and] had turned into hills of ruins. Assurnasirpal, king of Assyria, took [this city] in hand for renovation. I put a wall around it, founded therein a palace for my royal residence, [and] decorated it more splendidly than ever before. I stored therein barley [and] straw from all the [surrounding] land. I named it Dur-Assur.*

The local rulers of Zamua, as well as those from across the subjugated territories reaching from the Zagros to Syria, were required to supply labourers and materials for the Assyrian heartlands. Many of these resources were used by Assurnasirpal in the establishment of the site of Kalhu (modern Nimrud) as his royal residence. Located near the junction of the Tigris and Greater Zab rivers, the site was considerably better situated than the ancient capital of Assur as a place from which to launch campaigns to the north of Assyria. Work began in 879 BC when huge numbers of people were drafted in to undertake their required corvée work along with gangs of forced labourers. City walls were constructed that enclosed an area of 360 hectares, including the old, much-neglected settlement mound where temples and palaces were constructed from thousands of mudbricks. A canal was dug from the Greater Zab to the city where orchards and gardens were planted containing regional trees and plants as well as exotic examples gathered while the king was on campaign. An enormous mudbrick palace (the so-called North-West Palace) was the king's most magnificent construction. Massive blocks of gypsum were cut from nearby quarries and carved into gigantic, human-headed winged bulls and lions that were dragged to Kalhu and set up to guard the palace doorways. Other stone slabs lined the walls of important rooms and courtyards and were carved in relief with typical Assyrian royal images of the hunt, military campaigns, and rituals, accompanied by cuneiform inscriptions extolling Assurnasirpal's titles and triumphs. No earlier Assyrian sculptures could match these magnificent carvings in technical accomplishment. Some of the relief panels depict

Assurnasirpal's campaigns into the Zagros. One such example comes from the so-called West Wing of the palace (fig.64). It depicts the king travelling with attendants in a chariot over schematic mountains and spiralling rivers. Distinguished by his distinctive headdress, he holds a bow in his lowered left hand and two arrows in his raised right hand as the victorious warrior. Other reliefs from Kalhu were intended to establish a connection between the king and his gods. Images of supernatural spirits were erected at doorways and against the walls of ritually significant areas of the palace (fig.65). Known as an *apkallu*, this particular spirit was especially important as the relief was designed to provide magical protection for the royal throne room. It was

**FIG. 64** Gypsum relief panel showing Assurnasirpal II on campaign, from the North-West Palace, Nimrud, Iraq, *c.*875–860 BC, ht. 101 cm, 1856,0909.210 © Trustees of the British Museum

one of a pair of supernatural beings with the head of a bird that flanked an entrance together with a pair of colossal, human-headed winged lions. Originally highlighted with paint, he wears fringed robes and holds a cone and a bucket. The cone is described in Assyrian texts as a purifier and was used presumably to sprinkle liquid from the bucket. The spirit is thus shown in the act of purifying the gateway, helping to create a magical barrier repelling the forces of chaos.

Both Assurnasirpal and his son Shalmaneser III (858–824 BC) marched their armies regularly to the west where the states on either side of the Euphrates River provided Assyria with a source of tribute or plundered riches. To increase the efficiency of these campaigns, Shalmaneser created a number of administrative and supply centres, including the Aramaean city of Til-Barsip (modern Tell Ahmar) which he renamed as Kar-Shalmaneser. Having solidified his western front, the Assyrian king turned to confront Urartu which had emerged as a powerful state in eastern Anatolia centred on Lake Van. Urartu was beginning to threaten the supply of horses to Assyria, which were essential for maintaining its military dominance – at a time when cavalry was rapidly replacing chariotry as the main mobile force in most Near Eastern armies. In 856 BC Shalmaneser led his forces north, following the Tigris to the northern shores of Lake Van. This was formidably difficult terrain in which to campaign, with high mountains separating steep valleys. However, according to the Assyrian version of events, the forces of King Aramu of Urartu were defeated and Shalmaneser marched on to Lake Urmia, where he plundered the neighbouring towns and villages.

Assyria was the most powerful kingdom of the wider region and felt able to intervene even in the politics of its southern neighbour Babylonia. In 851 BC, the Babylonian king Marduk-zakir-shumi's position was challenged by his brother. In response the Assyrian army invaded the region of the Diyala and suppressed the rebellion. The relationship between Assyria and Babylonia was commemorated by a carved relief of the two kings shaking hands on a throne base in Shalmaneser's new palace (so-called Fort Shalmaneser) at Kalhu. The Assyrian king travelled to Babylon, Borsippa, and Cutha to present offerings at their temples. He was also enriched when his forces plundered the Sealand in the far south of Babylonia. Here three powerful Chaldaean tribes (Bit-Dakkuri, Bit-Amukani, and Bit-Yakin) had come to dominate the region and were becoming increasingly wealthy and powerful through their control of the Persian Gulf trade. Each tribe was centred on a fortified town where their king lived and although nominally subject to the ruler of Babylon they were essentially independent.

**FIG. 65** Gypsum relief panel with a protective spirit, from the North-West Palace, Nimrud, Iraq, *c.*875–860 BC, ht. 232 cm, AN1982.224 © Ashmolean Museum, University of Oxford

ABOVE:

**FIG. 66** Bronze horse-bit with cheek-pieces cast as winged sphinxes, *c.*950–650 BC, l. of cheek-piece 10.2 cm, AN1965.830 © Ashmolean Museum, University of Oxford

OPPOSITE:

**FIG. 67** Bronze disc-headed pin with grotesque face and animals, *c.*950–650 BC, l. 20.7 cm, AN1951.358 © Ashmolean Museum, University of Oxford

**FIG. 68** Bronze pin with 'mistress of animals', *c.*950–650 BC, ht. 12.8 cm, AN1965.799 © Ashmolean Museum, University of Oxford

**FIG. 69** Bronze 'standard' with confronted lions, *c.*950–650 BC, ht. 13 cm, AN1965.788 © Ashmolean Museum, University of Oxford

**FIG. 70** Bronze 'standard' with 'master of animals', *c.*950–650 BC, ht. 19.5 cm, AN1965.794 © Ashmolean Museum, University of Oxford

## THE BRONZES OF LURISTAN

To the east of Babylonia, the valleys of the central Zagros (modern Luristan) were home to a rich metalworking society that produced a highly distinctive range of bronze objects in the period approximately 950–650 BC. These were possibly intended largely for the graves of local elites from which the vast majority were plundered in modern times. However, a few tombs have been excavated, as has the site of Surkh Dum which was probably a religious shrine in which bronzes were deposited. The metalwork can be grouped into distinct object types comprising decorated whetstone handles, various pieces of horse-harness, elaborate pins, and delicate 'ritual' objects. Given the economic and political importance of the horses being bred on the plains of western Iran it is perhaps not surprising that the skilled metalworkers of Luristan produced exquisitely decorated horse trappings. Horse-bits, such as the one illustrated here, were placed in graves under the dead person's head (fig.66). This example has the cheek-pieces in the shape of a mythological sphinx-like animal – a creature known also from later Elamite glazed bricks – but other horse-bits depict horses, goats, bulls, or lions. Other elaborately decorated objects include pins which are among the masterpieces of Luristan metalwork. They were presumably used to fasten garments and in dressing hair. Here a disc-headed pin has a double ring of dots around a central boss in the form of a grotesque face framed

Fig. 67

Fig. 68

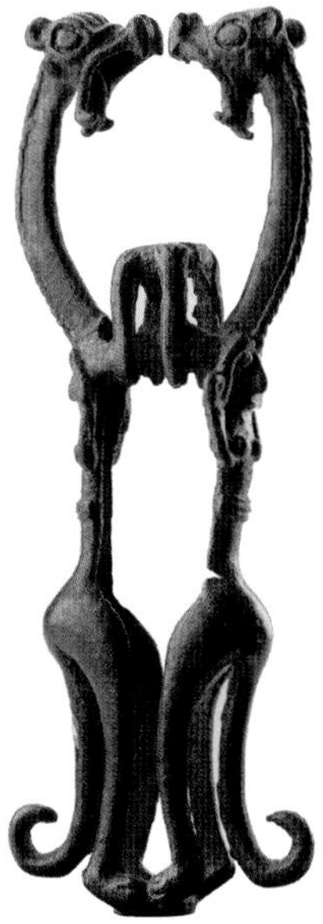

Fig. 69

Fig. 70

by a frieze of animals: antelope or gazelle, wild asses, lions, and a vulture (fig.67). Among the most popular form of votive pins is that represented by fig.68 which shows a central horned 'mistress of animals' holding inverted lions within a crescent frame terminating at the top in lion heads. The most distinctive objects from Luristan, however, are so-called finials or standards that appear to have been mounted on the top of hollow, bottle-shaped supports. Their purpose remains unknown and they are without parallel in the wider region. One type consists of a pair of freestanding rampant confronted animals, such as these two lions with wide, threatening jaws (fig.69). More elaborate and mysterious forms include fig.70 that is made up of two human faces on a central tube. The upper face takes the form of a 'master of animals', a figure symmetrically flanked by wild animals that he subdues by grasping their necks. The central face appears to 'wear' the upper figure as an elaborate headdress from which a pair of bird heads hangs. Some object may have been placed in the tube – a flower or slender branch perhaps.

## ASSYRIA, BABYLONIA AND URARTU

Assyrian campaigns in the upper Diyala and Zagros – including the transformation of Zamua into a directly governed province – as well as through Syria as far west as south-eastern Anatolia, meant that by 833 BC Shalmaneser III was able to march without resistance across a wide region to collect tribute from numerous kingdoms. However, in the last decade of his reign Sarduri I of Urartu emerged as a powerful opponent. He had established his capital at the rock-fortress of Tushpa on Lake Van from where he initiated the expansion of Urartu into north-western Iran. By incorporating territory through the mountains to the north of Assyria, Sarduri threatened Shalmaneser's hard-won tributary states. Between 832 and 827 BC five campaigns were launched from Assyria into the mountains to confront Urartian encroachment and ensure the continued flow of resources, especially of the all-important horses. Of particular concern was the fertile Ushnu-Solduz valley, a strategic east–west route south of Lake Urmia, within which was the kingdom of Mannaea.

Further attempts by Assyria to ensure political and commercial control over trade routes through north-western Iran were put on hold in the final years of Shalmaneser's reign when some of his own sons rebelled and engulfed the kingdom in civil war. Stability was finally restored by Shamshi-Adad V (823–811 BC), probably the legitimate heir, but only in around 820 BC, possibly with the support of the Babylonian king Marduk-zakir-shumi. Shamshi-Adad acted quickly to secure his northern borders and raids were conducted through the eastern mountains, including one against 'Sagbita, the royal city of Hanashiruka the Mede'. The Medes were Iranian speaking tribespeople and the city of Sagbita is almost certainly Ecbatana (modern Hamadan) that straddled the Great Khorasan Road on the eastern side of the Zagros.

The last years of Shamshi-Adad's reign were devoted to dealing with problems in Babylonia. Although Assyria and Babylonia had enjoyed a relatively harmonious relationship for much of the century, the new Babylonian king Marduk-balassu-iqbi must have turned against Assyria around 814 BC. Shamshi-Adad invaded the region east of the Tigris and besieged the Baylonian ruler at Dur-Papsukkal, a royal city near Der. This brought the Assyrian army into areas of traditional Elamite influence; they now re-emerge into history through the Assyrian royal inscriptions. Elamite troops, we are informed, joined with Kassite, Aramaean, and Chaldaean units, and went to the aid of Marduk-balassu-iqbi. Despite the united forces ranged against him, Shamshi-Adad captured the Babylonian king, plundered the country, and received tribute from the Chaldaean tribal leaders. The latter now began to fill the political vacuum left in Babylonia. When Shamshi-Adad's son, Adad-nirari III (810–783 BC) was crowned as the new Assyrian king, he also campaigned in Babylonia but there was no attempt to appoint his own ruler or claim the throne. Instead he contented himself with offering sacrifices in the temples of Babylon and Borsippa and receiving tribute from the Chaldaeans.

By the end of the ninth century BC, perhaps taking advantage of Assyria's focus on Babylonia, King Ishpuini of Urartu had advanced his borders to the western shores of Lake Urmia, threatening Mannaea to the south and its links with Assyria. His son, Menua (*c.*800–785 BC), having started the construction of a series of massive fortresses on the northern shore of Lake Van, annexed territories around the headwaters of the Tigris. The consequences of this for Assyria were profound and over the following century the kingdom's incursions into the Zagros to capture horses and other resources were transformed into campaigns for direct control of territory on the Iranian Plateau as well as confrontations with Urartu over control of trade routes, especially the Great Khorasan Road. Attempts by the two kingdoms to establish their influence among the urban centres and semi-mobile populations of north-western Iran inevitably led to conflict as these attempted to align with one power or the other. The dangers associated with this delicate political balancing act are perhaps evident in the Solduz Valley where an Urartian inscription at the site of Qalatgah records the establishment of Urartian power around 800 BC. Twenty kilometres to the east of this fortress lay the wealthy commercial centre of Hasanlu (its ancient name is unknown). Here excavations have revealed an impressive citadel that had been violently destroyed by fire sometime in the eighth century BC. The collapse of buildings buried, and preserved, many artefacts but also tragically many of the inhabitants: the skeletons of more than 240 men, women, and children were recovered from beneath the fallen walls and roofs. Four large baked-brick buildings had a common layout with an elaborate entryway and a central columned hall flanked by smaller rooms; they were probably elite residences or gathering places. The scale and quality of the buildings as well as the objects found in them suggest that Hasanlu was an important administrative and political centre. Many of the objects, made from ivory, glass, and precious metals, were imported from northern Syria and Assyria, reflecting Hasanlu's wide-ranging connections. Among the most spectacular objects recovered was a large gold beaker (fig.71). Its hammered decoration includes three gods riding in chariots and the scene may reflect regional myths.

The acute external pressure on Assyria in the first half of the eighth century BC is probably reflected in the kingdom's weakening ability to maintain a strong hold over its western territories – repeated campaigns were necessary to keep rebellious states in check. Tensions with Babylonia also began to emerge over the piedmont as political power shifted to Chaldaean tribes in the south; around 770 BC a leader of the Bit-Yakin tribe, Eriba-Marduk, claimed the Babylonian throne. Perhaps most tellingly, between 763 and 760 BC the Assyrian king faced rebellions at home in Assur and Arrapha. By the middle of the century, Urartian forces had advanced into the foothills of northern Syria and through the Zagros as far south as the Diyala River.

**FIG. 71** Gold bowl from Hasanlu, Iran, *c.*850–750 BC, ht. *c.*20 cm, © The Art Archive / Archaeological Museum Tehran / Gianni Dagli Orti

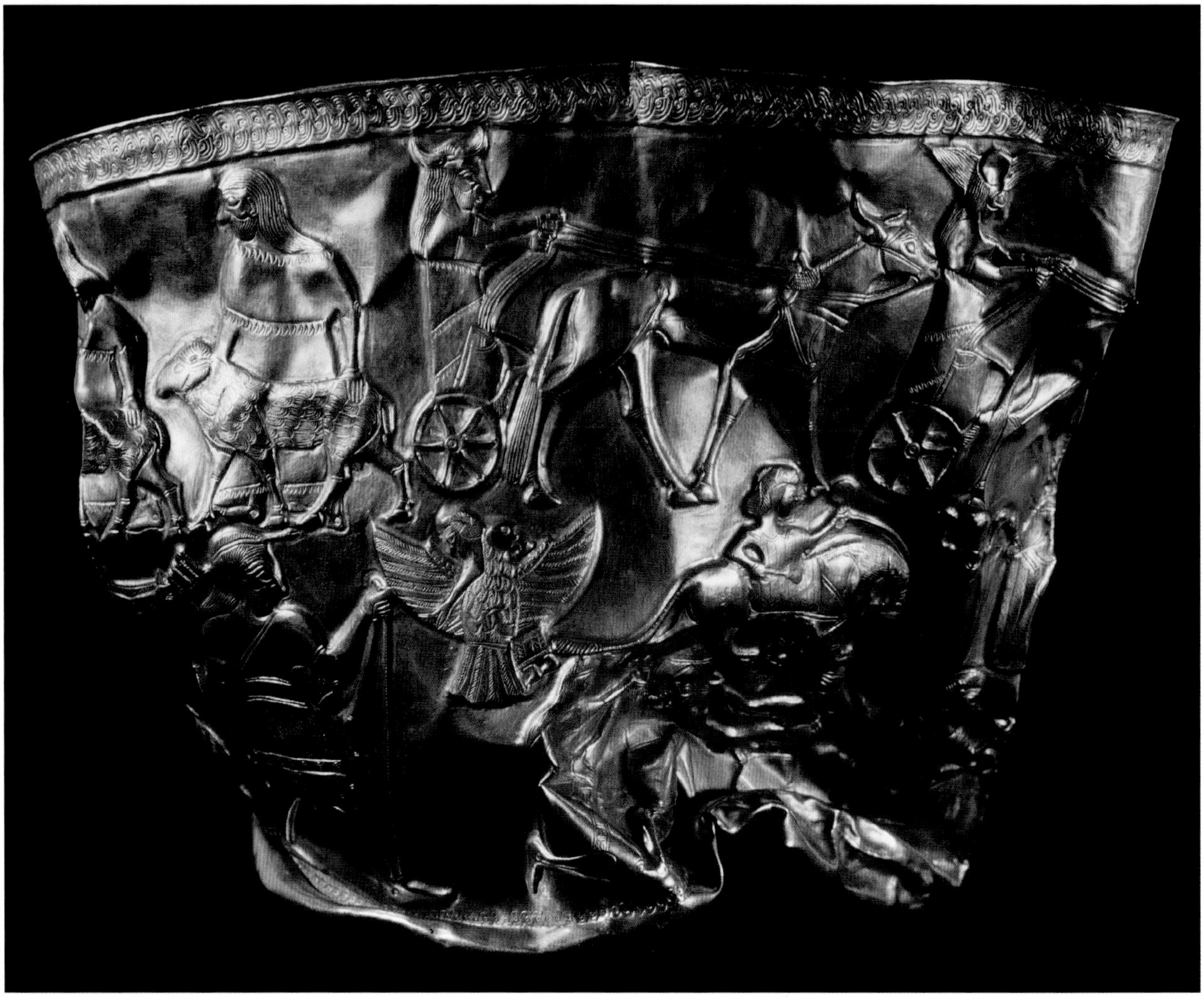

## ASSYRIA EAST OF THE ZAGROS

The dominance of Urartu and the lack of an effective response by Assyria were probably the underlying reasons for a rebellion at Kalhu which brought Tiglath-pileser III (744–727 BC) to the throne. He acted immediately to attack the root cause of Assyria's weakening position and in his first year Tiglath-pileser established two new administrative provinces within Iran: Parsua, located around the north-eastern headwaters of the Diyala valley and adjoining the province of Mazamua (the earlier kingdom of Zamua) that shared its northern border with the kingdom of Mannaea, and Bit-Hamban, to the south of Parsua along the main route of the Great Khorasan Road.

For the first time in its history a Mesopotamian kingdom directly controlled territory situated on the eastern slopes of the Zagros and along the Iranian part of the Great Khorasan Road. Assyria's new neighbours to the east were the Medes and in 737 BC the governor of Mazamua led campaigns 'against the mighty Medes of the rising sun', possibly as far east as Mount Damavand to the north-east of modern Tehran. The expansion of Assyrian authority onto the Iranian Plateau secured a supply of horses and metals for the kingdom's military needs but also inevitably drew it into the complex politics of the region within a challenging physical landscape. According to the Assyrian records there were small independent states both within and east of the Zagros, many of which they identified as Median. Most of them lay between the states of Mannaea in the north-west and Ellipi in the south-west. Although the Medes bred cattle and horses, their prosperity was probably largely dependent on controlling the various branches of the Great Khorasan Road, taxing traders, or demanding protection money. Each independent region was ruled by a 'city lord' from a fortified settlement. The best known sites that can possibly be described as Median, are Godin Tepe, Tepe Baba Jan, Tepe Nush-i Jan, and Tepe Ozbaki. As we have seen, Godin Tepe in the Kangavar valley, had already been the site of a trading centre and control point in the late fourth millennium BC. In the eighth or seventh century BC an impressive fortified citadel complex was constructed with three columned halls along with associated kitchen and storage areas; it was almost certainly the seat of a prominent Median 'city lord'. The style of pottery at Godin is similar to that of Tepe Baba Jan in Luristan. Here was a small fortified manor formed from two columned halls albeit much smaller than those at Godin. It may have been the residence of a minor political leader. Painted terracotta tiles that decorated the ceiling in one of the halls were preserved when they fell to the ground following a fire in around 700 BC (fig.72). On the eastern slopes of the Zagros, some 60 kilometres south

**FIG. 72** Painted terracotta tiles from Baba Jan, Iran, *c.*700 BC, ht. each tile approx. 40 cm, AN1968.1527–1529 © Ashmolean Museum, University of Oxford

of Ecbatana, lay Nush-i Jan. It consisted of a thick-walled fort with store rooms, a large hall with twelve columns, and a tower-like structure that can be interpreted as a fire sanctuary since it held a freestanding altar with a fire bowl, surrounded by other buildings including a second temple. Similar buildings were constructed at Tepe Ozbaki, about 75 kilometres north-west of modern Tehran, including a square columned hall.

The Medes would play an increasingly influential role in relation to the Assyrian presence in Iran but the most immediate pressing concern was the continued threat from Urartu. In 743 BC the Assyrian army overtook the forces of Sarduri II (about 765–735 BC) west of the Euphrates and captured the Urartian camp. Within ten years, Tiglath-pileser III had effectively contained Syria, and the region as far south as Damascus was reorganised into a series of provinces placed under the control of governors appointed directly by the Assyrian king. The expansion of the provincial system in both the east and west restored Assyria to a first-rank power and transformed the kingdom into an empire. The governors, supported by teams of scribal administrators, were responsible for the safe conduct of merchants and the maintenance of a road system equipped with way stations that stocked provisions for royal messengers and officials. The provincial administrators also managed the thousands of people who were forcibly moved as a result of military campaigns, and resettled in different areas of the empire where they were established as agricultural workers, set to work on building projects, or incorporated into the army.

Tiglath-pileser followed these achievements with a campaign into the east Tigris region, which was recaptured. The Assyrian army then moved further into Babylonia and received tribute from the Chaldaean tribes. Opportunities to extend his control further emerged when the Babylonian king, Nabu-nadin-zeri (733–732 BC), was assassinated by a local governor, who in turn was removed from power by the ruler of the Bit-Amukani Chaldaean tribe. Against this politically unstable country Tiglath-pileser launched a series of campaigns starting in 731 BC. After two years of exploiting the rivalry between cities and tribal groups, he was able to claim the throne of Babylon himself, a position which was cemented when he received acknowledgement and royal gifts from the leader of the wealthiest Chaldaean tribe, Marduk-apla-iddina II of Bit-Yakin.

Tiglath-pileser's son and successor Shalmaneser V (726–722 BC) extended Assyrian provinces in the west by capturing Samaria, capital of the kingdom of Israel, but may have been killed in a coup led by his brother, Sargon II (721–705 BC). The result of this political instability at the centre of Assyria led to rebellions among some of the remaining independent Syrian kingdoms, as well as among the tribal leaders of Babylonia. Yaubidi, the ruler of Hamath, led a coalition of neighbouring states in a bid to shake off the Assyrian presence in Syria, while the ruler of Gaza also rebelled, aided by the Egyptians. Sargon moved to crush the opposition; Yaubidi's coalition was defeated and both he and the king of Gaza were captured and flayed. Hamath was turned into an Assyrian province and control was re-imposed on Arpad, Samaria, and Damascus. In Babylonia, however, Sargon faced a far more daunting threat. In 720 BC, Marduk-apla-iddina of Bit-Yakin

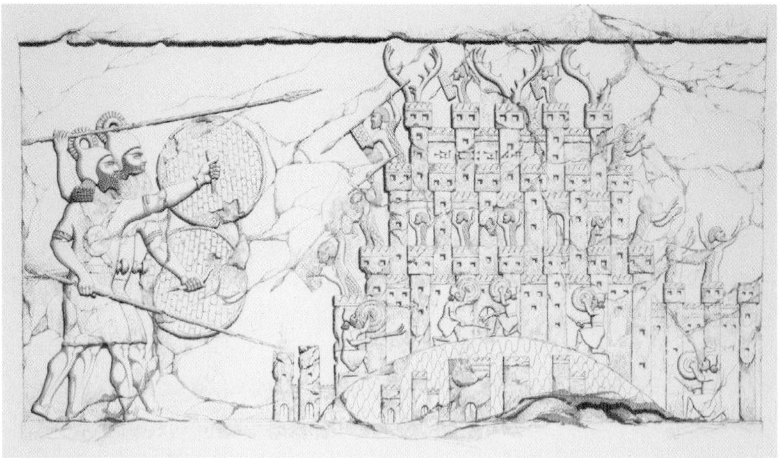

**FIG. 73** Drawing of a stone relief from the palace of Sargon II at Khorsabad, Iraq, showing the siege of Kishessim, from Botta, P.-É. and E. Flandin, *Monument de Ninive*, vol. I (Paris, 1849)

had allied himself with the Elamite king Humban-nikash (Ummanigash of the Babylonian texts) (743–717 BC), who, drawing on the manpower of the highlands, led his forces into Babylonia and met the Assyrian army in battle at Der. There was no clear outcome. Despite claiming a victory, it appears that Sargon was unable to capitalise on this but the Assyrians probably maintained their front line at Der. The crown of Babylon, however, was taken by Marduk-apla-iddina whose own inscriptions claim a victory. However, what may be a more impartial document – the so-called Babylonian Chronicle – credits this to the Elamites and even claims that the Chaldaean ruler did not arrive in time to enter into battle. Whatever the truth, the older connection between the rulers of Babylon and Elam was revived, but now with each kingdom attempting to use the other in their endeavours to outflank Assyria.

Despite his failure to hold the throne of Babylon, Sargon strengthened his position by establishing a number of new imperial provinces over the following years. In 717 BC the wealthy Syrian kingdom of Carchemish was conquered and turned into the seat of a provincial governor. Considerable plunder from this great trading city was carried to Kalhu and stored in the refurbished North-West Palace. There were also rebellions to deal with in the eastern Zagros. To the south of the Assyrian province of Parsua lay the small state of Harhar. The inhabitants overthrew their governor, ceased to pay tribute to Assyria, and formed an alliance with the mountainous kingdom of Ellipi further south. In 716 BC Sargon reconquered Harhar, which was renamed as Kar-Sharruken (Harbour of Sargon) and turned into the centre of a province, as was the Median fortress of Kishessim that was captured after a siege and renamed Kar-Nergal. The siege of Kishessim would be depicted on a wall relief in Sargon's palace at Dur-Sharruken (modern Khorsabad) (fig.73). The city is identified by an accompanying cuneiform inscription and shown with its battlements built upon a mountain and decorated with deer antlers. While some inhabitants are still fighting, others have abandoned their weapons and raise their arms in submission.

The Assyrian governors appointed to control these new Zagros provinces did not find this easy. The surviving correspondence between the king and his officials makes reference to continuous local resistance. In addition, the Assyrians had to contend with the extreme weather conditions in the mountains: heavy snowfall slowed building works and often cut off the provinces from Assyria. Over the following years the Assyrian army had to return regularly to crush rebellions resulting in the death and deportation of thousands of the inhabitants. A number of these problems faced by the Assyrians may have resulted from the decision to leave the local 'city lords' in power. However, after several years the troubles subsided, presumably because the 'city lords' and the provincial administrators reached mutually advantageous and profitable arrangements. Meanwhile, the people who had been deported from their homelands did their best to settle in new lands and to adjust to what in many cases must have been a brutal change in their way of life. Some families, however, seemed to have fared well as suggested by the contents of an archive of tablets found in a large house in the city of Assur. Dating to the later seventh century BC it records the business activities of a prosperous family who had originated in Hundir, close to the Median fortress of Kishessim, and may have arrived in Assyria as deportees from Sargon's crushing of that rebellious city in 716 BC or soon after. Several generations later, the family had been fully integrated into Assyria and were producing a certain type of rug, perhaps to be understood as hand-knotted carpets. The family was also involved in long-distance overland trade and were probably making full use of connections with their former homeland.

Meanwhile in Assyria the king had been resisting Urartian attempts under Rusa I to push south through the kingdom of Mannaea. In 714 BC Sargon marched through Mannaea and received the submission of the western Medes. In a letter written in high literary language and addressed to the god Assur, Sargon recounts the campaign:

> *Mount Simirria, a great mountain peak that points upwards like the blade of a lance, and raises its head over the mountain where the goddess Belet-ili lives, whose two peaks lean against heaven on high, whose foundations reach into the midst of the netherworld below, which like the back of a fish, has no road from one side to the other and whose ascent is difficult from the front or back, ravines and chasms are deeply cut in its side, and seen from afar, it is shrouded in fear, it is not a good climb in a chariot or with galloping horses, and it is very hard to make infantry progress in it; yet, with the intelligence and wisdom that the gods Ea and Belet-ili destined for me and who broadened my stride to level the enemy land, I made my engineers carry heavy bronze axes, and they*

*smashed the peaks of the high mountain as if it were limestone and made the road smooth. I took the head of my army and made the chariots, cavalry, and battle troops that accompany me fly over it like eagles. I made the support troops and foot soldiers follow them, and the camels and pack mules jumped over the peaks like goats raised in the mountains. I made the surging flood of Assyrians easily cross over its difficult height and on top of that mountain I set up camp.*

Moving through the rugged, forested terrain of north-western Iran, Sargon's army surprised the Urartian forces along with their Mannaean allies in a night attack from which the Urartian king Rusa I (*c.*734-714 BC) only just managed to escape. From the western shores of Lake Urmia, Sargon turned south and fell upon the small state of Musasir whose ruler had allegedly reneged on his agreements with Assyria. The city was captured and the treasuries of the palace and temple plundered; some 335,500 objects and more than 100,000 kilograms of copper were taken. Possibly contributing to the Assyrian successes was pressure mounting on the northern borders of Urartu from Cimmerian nomadic raiders who were moving through the Caucasus Mountains and along the shore of the Black Sea. Rusa gathered his forces against the Cimmerians but he was either killed or, according to the Assyrians, committed suicide following his defeat and was succeeded by his son Argishti II.

## ATTEMPTS TO DOMINATE BABYLONIA AND IRAN

By 710 BC substantial opposition to Assyrian authority had been crushed and Sargon was able to confront Marduk-apla-iddina who had claimed the Babylonian crown for the previous ten years. There was always the danger that the Babylonians would ally themselves with the Elamites, resulting in a repeat performance of the battle of Der. Sargon therefore cautiously avoided a direct attack on Babylon and skillfully dealt with Marduk-apla-iddina's Aramaean tribal allies (including deporting up to 18,430 people from the city of Dur-Athara). Although the Chaldaean leader sent a bribe (a bed, throne, footstool, a silver water basin, and jewellery) to the new Elamite king Shutruk-Nahhunte II (717–699 BC) no support was forthcoming. Without Elamite opposition Sargon entered Babylon – at the invitation of its representatives according to his inscriptions – and was crowned king. The following year the Assyrians prepared to attack Marduk-apla-iddina, who had withdrawn to his homeland of Bit-Yakin. His tribal capital Dur-Yakin was heavily defended but Sargon captured it and seized huge quantities of

FIG. 74 Reliefs from the palace of Sargon II at Khorsabad, Iraq, showing Medes bringing tribute, *c.*710–705 BC, l. 307 cm, AO19887 © RMN-Grand Palais (musée du Louvre) / Hervé Lewandowski

property. Marduk-apla-idinna was injured in the hand by an arrow, but he managed to escape, though the location of his refuge is unknown.

Bit-Yakin had grown wealthy through its trading relations with the Persian Gulf and as the terminus for the main caravan route running across the desert from southern Arabia. Now the king of Dilmun (Bahrain) sent an embassy of friendship to the Assyrian king. Sargon's wealth and status was mirrored by the enormous royal centre he had constructed some nineteen kilometres north-east of Nineveh on land he purchased from local communities. Surrounded by a wall seven kilometres long, the city was named Dur-Sharruken ('Fortress of Sargon', modern Khorsabad) and contained a royal palace and temples built on an artificial terrace. The court moved to the new capital in 707 BC even though the construction was not completely finished. The palace consisted of as many as 240 rooms, many of which were decorated with colossal stone panels carved in relief with scenes of battles, sieges, hunting, and feasting. The walls of courtyards and many rooms in the palace were decorated with reliefs showing formal processions of larger-than-life courtiers and tribute-bearers approaching the king. Such images represented an ordered world established by Sargon in which tribute came to him from east and west. Among them are Medes bringing horses and other gifts. Many Median rulers established treaties with the Assyrian king so that, as vassals, they were protected against Assyria itself but also against other states such as Mannaea and Urartu. In return Assyria received regular tribute in the form of horses – unlike Assyria's vassals elsewhere whose tribute was calculated in metal. This relief from Dur-Sharruken shows three Medes leading two stallions (fig.74). The individual at the head of the procession holds in his left hand a city model as a symbol of his country and raises his right hand in a gesture of respect. The men wear animal skins as part of their clothing, a Mesopotamian artistic convention since the third millennium BC for depicting eastern mountain people (see fig.45).

Creating and maintaining such a vast empire had required Sargon to undertake repeated military campaigns. Instability could be expected beyond the borders of the distant provinces and in 705 BC Sargon found himself faced with a serious threat. It is unclear where the trouble erupted, but it was probably connected with Cimmerian nomadic incursions into north-western Iran and eastern Anatolia. Sargon, at the head of his forces, was killed and the whole Assyrian camp was captured.

This was potentially disastrous for Assyria since the death of a king could lead to rebellions. Indeed the Babylonian throne was claimed by a certain Marduk-zakir-shumi. Nonetheless, Sargon's son and heir Sennacherib (704–681 BC) took the throne without any known difficulties. The new king relocated the capital and its administration to Nineveh where he initiated the transformation of that city into a magnificent metropolis. The building project would take around ten years to complete. At its heart lay the ancient settlement mount on which was constructed an impressive palace (known today as the South-West Palace). Nearly every room in the building was decorated with wall reliefs. These largely record details of Sennacherib's first three military campaigns along with scenes depicting the building of Nineveh itself. The city was surrounded by agricultural and park lands supplied with water brought over 65 kilometres by canals and aqueducts from the Zagros Mountains.

Sennacherib's first serious challenge was to re-establish Assyria's hold on Babylonia. In 703 BC Marduk-zakir-shumi had been deposed by Marduk-apla-iddina who, at the mobilisation of Assyrian forces, called upon the support of Aramaean and Chaldaean tribes but, most importantly, again sought help from the Elamites. According to Sennacherib's account, the Babylonian king turned to Shutruk-Nahhunte II and gave him gold, silver, and precious stones. On this occasion, for unknown reasons, the Elamite king agreed to support him and sent 80,000 bowmen – a key component of an Elamite army. However, as the Assyrians entered Babylonia, Marduk-apla-iddina withdrew to his tribal territory in the south of Babylonia, leaving the Elamite forces to be driven back by the Assyrian army. Sennacherib entered Babylon where he appointed as his representative a man called Bel-ibni, described by the king as having been brought up at the Assyrian court 'like a puppy'.

The following year Sennacherib led an assault by the Assyrian army on territories along the Great Khorasan Road into the Zagros Mountains. The campaign continued with a march south through the kingdom of Ellipi, which was reinforced as a buffer zone between the province of Harhar and Elam.

The expedition ended with the receipt of tribute from the Medes and many of them may also have been deported to other areas of the Assyrian Empire.

In 701 BC the Assyrian king focused on re-establishing his authority along the eastern Mediterranean coast where a significant rebellion had broken out among Phoenician, Philistine, and Hebrew kingdoms, conspiring with Egypt. Nonetheless, once order had been restored in the west, it would be relations with Babylonia and Elam that would remain Sennacherib's principal concern for the rest of his reign. Indeed by 700 BC Bel-ibni was intriguing with anti-Assyrian groups and Mushezib-Marduk, the ruler of the Chaldaean Bit-Dakkuri tribe, had gathered his forces in opposition to Assyria. The Assyrian army marched into southern Babylonia where it crushed the opposition and Mushezib-Marduk fled. As Sennacherib advanced into Bit-Yakin territory, Marduk-apla-iddina sought refuge in Elamite territory by crossing the Persian Gulf and, according to Assyrian royal inscriptions, he 'disappeared'. Bit-Yakin was devastated by the Assyrians, who re-imposed their authority in Babylon; Bel-ibni was captured and taken back to Assyria. It may be this campaign that is depicted on wall panels from Sennacherib's new palace at Nineveh. This fragment showing a captive woman and child riding on a mule being escorted by Assyrian soldiers was probably part of a larger scene depicting the deportation of peoples from Babylonia, which is identified by the presence of fruiting date palms (fig.75).

**FIG. 75** Gypsum relief panel showing the deportation of Babylonian prisoners, from the South-West Palace, Nineveh, Iraq, *c.*700–695 BC, ht. 92cm, AN1933.575 © Ashmolean Museum, University of Oxford

Sennacherib appointed his son and heir, Assur-nadin-shumi (699–694 BC), as the ruler of Babylon and for six years the country appears to have enjoyed a period of stability. It is evident from this monument, which is the only known portrait of him, that Assur-nadin-shumi presented himself as a traditional Babylonian ruler (fig.76). His image is carved on a so-called *kudurru* (boundary-stone) or *naru* (standing stone), a long-established object in Babylonia used to record royal grants to high-ranking individuals of land, tax exemptions, and privileges. Assur-nadin-shumi wears the crown of Babylon and is named in the cuneiform inscription that covers the back and front of the monument. Above the king is a so-called *mushhushu* snake-dragon, the sacred animal of several important deities including Marduk, the supreme god of Babylon.

In 694 BC Sennacherib felt the need to conquer the people of the Bit Yakin tribe who had followed Marduk-apla-iddina across the northern Persian Gulf into Elamite territory. Ships were built in Nineveh by Syrian craftsmen and manned by Phoenicians that Sennacherib had previously deported from the west. They were floated down the Tigris to the city of Opis in northern Babylonia (the exact location of the site is unknown) from where they were dragged overland to a branch of the Euphrates River. At the Persian Gulf they waited for the sea to calm before the expedition set sail around the coast and, once ashore, took the people of Bit-Yakin captive and shipped them back to Assyria. Meanwhile, however, the Elamite king, Hallushu-Inshushinak (699–693 BC) had taken full advantage of the Assyrian army's presence at the coast to attack northern Babylonia, perhaps via the Diyala. He moved his army on to the plains around Sippar and captured both the city and Assur-nadin-shumi, who had been betrayed to him by some Babylonian nobles. Sennacherib's son was taken captive to Elam and Hallushu-Inshushinak's forces successfully fought off a counter attack by a contingent of Assyrian troops. With Elamite support a certain Nergal-Ushezib claimed the Babylonian throne and extended his influence to include the cities of Nippur and Uruk. Sennacherib fought his way north and a fierce struggle continued into the following year. Eventually a crucial battle was fought near Nippur and Nergal-Ushezib was captured.

The Assyrians now turned against Elam and a number of border cities were sacked and placed under the governor of Der. Perhaps as a result of this, the Elamites deposed Hallushu-Inshushinak in favour of Kudur-Nahhunte, but the new king fled into the mountains as the Assyrians advanced towards his royal city of Madaktu (perhaps modern Tepe Patak, northwest of Susa in the Deh Luran Plain). The approach of winter brought further Assyrian advances to a halt. Although Sennacherib had managed to

FIG. 76 Fragment of a black stone *kudurru* of Assur-nadin-shumi, *c.*699–694 BC, ht. 16 cm, AN1933.1101 © Ashmolean Museum, University of Oxford reclaim much Babylonian territory he had been unable to regain control of the north-west of the country, including the city of Babylon, and Mushezib-Marduk returned to claim the throne.

Another rebellion in Elam in 692 BC resulted in the death of Kudur-Nahhunte and he was replaced by Humban-nimena III (692–689 BC). The new Elamite king was persuaded by a considerable gift of gold and silver to offer military support to Mushezib-Marduk. The combined Elamite–Babylonian army included Aramaeans and Chaldaeans, and a coalition of highlanders including mercenaries from Parsua, Ellipi, and Anshan. The following year this army began to push north, following the River Tigris. At the site of Halule (which has yet to be identified) it confronted the Assyrian army and engaged in a ferocious battle. There are conflicting accounts of the outcome, with Babylonian texts describing an Assyrian defeat, while Sennacherib claims to have been victorious after killing 150,000 warriors. It is clear, however, that the Assyrians did not suffer any serious reverse because in 690 BC Sennacherib placed the city of Babylon itself under siege. For fifteen months the besieged people of Babylon suffered increasingly terrible conditions of famine and starvation as Mushezib-Marduk's allies gradually abandoned him or were defeated by the Assyrians. In 689 BC Humban-nimena III died following a stroke and in the same year Babylon fell to the besieging forces. Sennacherib took revenge on the rebellious population; the symbols of the city's wealth and cultural prestige, the palaces and temples, were looted and ruined. The great religious cults, especially that of the god Marduk, were suspended, a devastating loss both psychologically and economically for the city. Mushezib-Marduk was taken captive to Assyria and north-western Babylonia was abandoned to poverty. Other areas of Babylonia however, particularly the south, with their important trade connections were not ignored for long. In 681 BC Sennacherib returned divine statues to Uruk and governors were appointed to the regions of the Sealand and the city of Ur.

With the abduction and probable death of Assur-nadin-shumi at the hands of the Elamites, Sennacherib had chosen his younger son Esarhaddon to be crown prince. Esarhaddon's appointment was presumably the cause of jealousy among his older brothers since, as stated tersely by the Babylonian Chronicle: 'on the twentieth day of the month of Tebet [December/January] Sennacherib, king of Assyria, was killed by his son in a rebellion'. This was probably at the hands of his eldest son Arda-Mulissi, although Esarhaddon, who was absent from the royal court, might have engineered Sennacherib's death if his status as heir was being challenged. Civil war broke out and continued for over forty days as rival forces gathered around Esarhaddon and his brothers and their armies clashed in eastern Syria. By early 680 BC Esarhaddon (680–669 BC) had emerged with sufficient support from within Assyria to cause his brothers to flee, possibly to Urartu, leaving him to enter Nineveh and claim the throne.

Among the first challenges for the new Assyrian king were problems in the west – in 679 BC an Assyrian army marched to the Egyptian border and captured the city of Arza – and the north-east where it was necessary to repel an incursion of Cimmerians and Scythians pushing south from the Caucasus. Some of these mobile tribes moved into northern Iran while others headed into Anatolia: Medes, Mannaeans, and Cimmerians were accused in Assyrian reports of plundering border areas and capturing several fortresses. There were also disturbances in southern Babylonia, where ambitious tribal leaders, officials, and nobles continued to seek political advantage. Nabu-zer-kitti-lishir, the governor of Uruk – who happened to be a son of Marduk-apla-iddina – attempted to extend his authority over Ur and lay siege to the city. Esarhaddon mobilised Assyrian troops in the Sealand and Nabu-zer-kitti-lishir fled to Elam seeking sanctuary. The Elamite king, Humban-haltash II (?–675 BC) clearly had other plans and executed him. Meanwhile, Nabu-zer-kitti-lishir's brother had submitted to Esarhaddon, who installed him as governor of the Sealand. This brought a certain amount of political stability to the region although there are indications of further struggles for power between the remaining sons of Marduk-apla-iddina. Esarhaddon also developed a policy aimed at winning him additional support from the cities of northern Babylonia by presenting himself as a traditional ruler of the country. Gradually he introduced the idea of restoring the city of Babylon to its pre-eminent cultural and commercial position. Cult-statues were returned to various Babylonian cities and repairs to the damaged temples were undertaken.

Events in the Zagros and further east also demanded attention but the written records – Assyrian royal inscriptions, letters, and requests by the king's specialist diviners for answers from oracles – present a confusing picture. One oracle makes reference to Bartatua, king of the Scythians, who had requested marriage to an Assyrian princess. Other reports suggest that there may have been an Assyrian victory over a joint Mannaean–Scythian army in 676 BC. Campaigns were launched onto the Iranian Plateau and Assyrian troops reached the region of Mount Bikni (Mount Damavand, north of modern Tehran or perhaps Mount Elvend closer to Assyria) as they had under Tiglath-pileser III. The main aim appears to have been to stabilise the region and maintain tributary relations, as well as to secure horses. Esarhaddon claims that:

*Uppis, city ruler of Partakka, Zanasana, city ruler of Partukku and Ramateya, city ruler of Urkazabarna, distant-dwelling Medes, who in the time of the kings my fathers had never crossed over the frontier of Assyria nor set foot on its soil, terror of the majesty of Assur, my lord, overcame them. Choice great warhorses and blocks of lapis lazuli quarried in the mountain they brought to Nineveh, city of my lordship and kissed my feet.*

*Patusharri, a land bordering the salt desert, in the midst of the land of the distant Medes, by Bikni, the lapis lazuli mountain, on whose soil none of the kings my fathers had ever set foot – Shidirparna and Eparna, mighty city rulers, who had never submitted to a yoke, I deported to Assyria together with their people, their riding horses, cattle, small cattle, and camels, a heavy booty.*

Political stability was achieved elsewhere by diplomacy. In 675 BC the Elamite king Humban-haltash II died and his brother and successor Urtak (675–664 BC) established friendly relations with Esarhaddon through a treaty (*adê*) in which oaths were sworn to both Assyrian and Elamite gods. Of even greater significance, however, was an immensely long *adê*-oath sworn by all of Assyria's vassals and subjects designed to safeguard the accession of Esarhaddon's two sons to the thrones of Assyria and Babylonia. This took place in 672 BC, probably over several days during elaborate public ceremonies. Copies of the succession treaty were inscribed in Akkadian cuneiform on large clay tablets, possibly for public display in temples throughout the empire. A description on the tablet states that it is: 'The treaty which Esarhaddon, king of Assyria, concluded on behalf of Assurbanipal, the great crown prince designate of Assyria, and Shamash-shumu-ukin, crown prince designate of Babylon'. Some of the surviving copies – the tablets were discovered smashed into hundreds of fragments at Nimrud – were the oaths imposed on the tribes and kingdoms to the east of Assyria. Illustrated here in fig.77 is the top of the tablet recording the treaty between Esarhaddon and Humbaresh, ruler of the city of Nashimarta in Media. Three seals, described in the text as belonging to the god Assur, were impressed on all of the treaty tablets. The impression on the right is from an old seal dating to the reign of Tukulti-Ninurta I (1233–1197 BC). The kneeling king is presented to a god standing on a bull, probably Assur, while the storm god Adad looks on. The left-most impression is that of Sennacherib which is described in the accompanying text as Assur's 'Seal of Destinies' and 'Whatever he seals is not to be changed'. The seal thus

transformed the treaty into a divinely sanctioned document equivalent to the 'Tablet of Destinies' on which the fate of the cosmos was inscribed.

With the future secured, Esarhaddon turned his attention to the problems along the eastern Mediterranean coast where some of the wealthy Phoenician cities were strengthening their ties with Egypt. An Assyrian army had marched into Egypt in 674 BC but was driven back. Esarhaddon tried again in 671 BC when the Assyrians fought as far as the capital Memphis which was captured. As soon as the Assyrian army withdrew, however, Taharqa (*c.*690–664 BC), the king of Kush (modern Sudan), reclaimed the throne. In 669 BC Esarhaddon set out again for Egypt but died en route and the expedition was aborted.

FIG. 77 Fragmentary baked clay cuneiform treaty between Esarhaddon and Humbaresh with seal impressions, 672 BC, w. 28 cm, 1959,0414.8 © Trustees of the British Museum

## ASSYRIA TRIUMPHANT AND COLLAPSE

In Assyria Assurbanipal (668–*c.*631 BC) ascended the throne but it would not be until the following year that his brother Shamash-shum-ukin would be crowned in Babylon; indications were that the Babylonian ruler would play a subordinate role while real authority lay in Nineveh. Early problems for the Assyrian king lay in the west where a series of extensive military campaigns into Egypt in 667 and 664 BC removed Kushite authority over the Nile

FIG. 78 Detail of a limestone relief panel showing the defeat of the Elamite king Te'umman, from the South-West Palace, Nineveh, Iraq, *c.*650 BC, ht. of entire slab 173 cm, 1851,0902.8b © Trustees of the British Museum

Delta. In the east, friendly relations were maintained with Elam and, indeed, Assurbanipal provided aid when the region suffered from famine as the result of a prolonged drought. However, in 664 BC the Elamite king Urtak was persuaded by both the governor of Nippur and the leader of the Babylonian Gambulu tribe to invade Babylonia. Perhaps because the greater part of the Assyrian army was occupied in Egypt, the Assyrians delayed sending troops, but when they were at last mobilised the Elamites simply withdrew and there were no repercussions.

Nonetheless, when Urtak died later that year, the fractured nature of Elamite politics was revealed with the appearance of rival claimants for his throne. Eventually Urtak's family as well as that of his predecessor Humban-haltash II sought refuge in Assyria, while the crown of Elam was claimed by a certain Te'umman. Te'umman repeatedly demanded in letters that Assurbanipal should return the political refugees to Elam – according to Assyrian accounts these letters were extremely insulting. By 653 BC diplomacy had broken down and an Assyrian army attacked the rebellious Gambulu tribe and invaded Elam. In a great battle at Til-Tuba on the River Ulai (perhaps the river Karkheh) near Susa, Te'umman and his son were killed. Two of the refugee Elamite princes were installed (in the cities of Madaktu and Hidalu) as client rulers over parts of Elam. The Assyrian triumph was depicted in a magnificent series of stone reliefs in the South-West Palace at Nineveh. The death of Te'umman is portrayed as a continuous narrative: he and his son fall from their chariot and flee into the woods but are captured, killed, and beheaded (fig.78). The king's head was carried back to Assyria where, as also described in elaborate royal inscriptions, it was suspended from the Gambulu chief's neck before Assurbanipal personally mutilated it.

One of the newly appointed Elamite rulers, Ummanigash, proved to be a treacherous choice as he rebelled against Assyria almost immediately. He had a powerful, and not altogether surprising, ally in Assurbanipal's own brother Shamash-shum-ukin. The Babylonian king was almost certainly resentful that real authority lay at Nineveh but, after fifteen years on the throne, he had also probably become enmeshed in the complex politics of Babylonia with its pro- and anti-Assyrian factions. Shamash-shum-ukin had clearly been able to forge a meaningful power base and could rely on support from the cities of northern and central Babylonia, as well as many of the Chaldaean and Aramaean tribal areas and Arab groups. We are, as usual, reliant on the Assyrian royal accounts to reconstruct most of what took place. In 652 BC Elamite troops joined the Babylonian forces and fought a battle with the Assyrian army in northern Babylonia, perhaps in the Diyala region, where the allied army was defeated. The Assyrians rapidly forced Shamash-shum-ukin to withdraw into Babylon. Nonetheless, there remained fierce resistance to the Assyrians. By 650 BC Babylon was under siege, as were the cities of Borsippa, Cutha, and Sippar. Meanwhile in Elam, Ummanigash was murdered and replaced by Tammaritu but, while he too supported Shamash-shum-ukin with troops, he was himself overthrown by Indabibi. There are indications that Indabibi sought a treaty with Assurbanipal but failed to honour Assyrian demands. As an Assyrian army moved against him the Elamites revolted against Indabibi whom they killed and replaced with Ummanaldash.

After two years of siege in which the desperate inhabitants were reduced to cannibalism, the city of Babylon fell to the Assyrians in 648 BC and Shamash-shum-ukin died during the onslaught. Assyrian rule was re-imposed under an Assyrian appointee called Kandalanu (647–627 BC) which perhaps brought a level of stability to the country. This was not the case in Elam where local rulers vied for the royal throne. By 647 BC internal revolts had further weakened Elam and Assurbanipal took advantage of this by invading the region. Ummanaldash fled to the mountains as the Assyrian army laid waste to the towns and cities of Susiana. At Susa Assurbanipal plundered the Elamite treasure houses from which prodigious amounts of goods were sent back to Assyria. The great ziggurat at Susa was destroyed and the Assyrian royal inscriptions describe how the great bronze horns – symbols of divinity – that decorated the structure were broken off. Assurbanipal's revenge even included the desecration of tombs belonging to the ancestors of the Elamite kings. Local rulers now sent embassies to the Assyrian king. These included Kurash, king of Parsumash, whom some scholars have suggested might be the Persian king of Anshan, Cyrus I.

Assurbanipal marked the successes of his army with impressive celebrations. A new palace was constructed at Nineveh (the North Palace) in which the king had his victories over Egypt, Babylonia, and Elam depicted in impressive wall reliefs. Among them were images of the capture by Assyrian soldiers of Ummanaldash, and the humiliation of a number of Elamite rulers – wearing distinctive bulbous headdresses – at the hands of Assyrian courtiers with flywhisks at Nineveh (fig.79). Although the images evoked a vigorous and powerful monarch, this was perhaps in contrast to reality. Years of military campaigns had severely drained Assyria's resources and with the burden of supporting a huge bureaucracy there would be no further expensive large-scale military campaigns to maintain the sprawling empire, which had begun to fray at the edges.

As late as 643 BC the Urartian king Sarduri sent an embassy of friendship to Assurbanipal but within a short time, perhaps only a few years, the kingdom of Urartu had broken apart into smaller kingdoms and tribal lands as Cimmerians, Medes, and other groups captured territory across eastern Anatolia and north-western Iran. It was during this period that the various Median territories, especially those along the Great Khorasan Road, may have begun to forge much closer ties. The Assyrian Empire was the principal market for the products and raw materials that flowed along the trade routes that the Medes controlled, and as the empire began to weaken so did demand. The result may have been the development of a loose form of centralised authority focused on the Median centre of Ecbatana. Within decades this confederation would take full advantage of Assyria's decline.

The last dated document of Assurbanipal is 631 BC and he may have died that year. He was succeeded by his son Assur-etel-ilani (*c.*630–627 BC). However, the Assyrian throne was contested and for a brief period a non-royal official called Sin-shum-lishir was able to claim it. Eventually, another of Assurbanipal's sons, Sin-shar-ishkun (626–612 BC) became king but by

**FIG. 79** Gypsum relief panel showing Elamite kings being humiliated by Assyrian officials, from the North Palace, Nineveh, Iraq, *c.*645–635 BC, w. 68.6 cm, 1856,0909.55 © Trustees of the British Museum

this time Assyria was losing its hold on Babylonia. Kandalanu had died in 627 BC and new claimants for the Babylonian throne had emerged. Among the powerful individuals who were vying for control was a man called Nabopolassar who had gathered military backing from among some cities and tribal groups. Little is known of his background although it has been suggested that he may have been appointed as a general in Babylonia by the Assyrians and had now decided to make a bid for power. Initially the Assyrians were able to keep his forces at bay, but in 626 BC the Assyrian army suffered a reverse at Babylon and Nabopolassar claimed the throne. Babylonia continued to be divided between pro- and anti-Assyrian groups but gradually Nabopolassar extended his authority and by 616 BC he began to launch attacks on Assyrian territory. Assyria resisted fiercely but after many months Nabopolassar had pushed north along the Euphrates into Syria. The following year the Babylonians attacked into Arrapha but the Assyrians managed to repulse them at Assur.

With Nabopolassar's campaigns diverting Assyrian resources to the defence of the homeland, the Medes saw an opportunity to profit. The ruler of Ecbatana known to the Babylonians as Umakishtar – and to the later Greeks as Cyaxares – mobilised the Median forces and, probably taking the traditional route through the Zagros, advanced onto the Assyria plains in the late summer of 614 BC, occupied the province of Arrapha and attacked Nineveh. The Assyrian capital was well defended and the Medes were driven off, but they laid siege to Assur, which they captured and plundered.

Despite this setback, the Assyrian army was able to take the offensive in 613 BC and moved against Nabopolassar's forces. The strength of Assyria may have forced Nabopolassar to follow the tradition of earlier Babylonian kings and forge an alliance with eastern peoples. There is little evidence that Elam was politically unified or had the capacity to organise an effective army at this time. However, the Medes had clearly demonstrated their military capabilities and it was with Cyaxares that Nabopolassar formed a treaty (according to a later tradition this was sealed through the marriage of a daughter of Cyaxares with Nabopolassar's son, Nebuchadnezzar). The two armies joined forces and in 612 BC they began to besiege Nineveh. For three months the great city resisted. When Nineveh eventually fell, the palaces were plundered and severely damaged; Sin-shar-ishkun was probably killed. Yet the extraordinary strength of the Assyrian imperial system was demonstrated by the fact that it did not collapse. Local governors continued to resist the Babylonian and Median forces and an Assyrian army rallied at Harran in southern Anatolia. Here it was joined by an expeditionary force from Egypt, keen to extend its own power in the region.

In 610 BC the Babylonian and Median armies once again joined forces and advanced against Harran. The Assyrian and Egyptians avoided a siege by withdrawing from the city which Nabopolassar now plundered and garrisoned. The following year the Egyptian and Assyrian forces attempted to besiege Harran but the city did not fall. Assyria as a political entity was swept aside as the following decades witnessed a violent struggle between Egyptian and Babylonian armies for domination of the eastern Mediterranean.

## THE NEO-BABYLONIAN EMPIRE

Nabopolassar had initiated building works on a grand scale in many of the major cities throughout Babylonia. These were continued by his son and successor Nebuchadnezzar II (604–562 BC) who had been leading military campaigns in Syria and would go on to extend Babylonian control through the Lebanon Mountains and the kingdom of Judah. It was the restoration and beautification of his capital that most clearly reflected Nebuchadnezzar's power and wealth. Babylon was developed into a city that would rival Nineveh for its legendary magnificence. Covering an area of at least 850

**FIG. 80** Baked clay statuette (restored) of the god Nin-shubur from Kish, Iraq, *c.*580 BC, ht. 17.7 cm, AN1928.527 © Ashmolean Museum, University of Oxford

hectares, making it one of the largest settlements of the ancient world, the site was enclosed by colossal double walls and a moat. The Euphrates River flowed through the centre of the city and access by boat was controlled by metal gates. A stone bridge spanned the river and connected the western half of the city with the heart of the settlement on the eastern bank. Here lay the royal palaces and the temple of Marduk with its associated ziggurat. Other ancient cities of Babylonia such as Ur and Uruk were revived with wide-ranging building programmes, as was Kish, the settlement most closely associated in Mesopotamian tradition with kingship. This baked clay statuette was found in a brick box beneath the floor of a room in a temple built by Nebuchadnezzar on a part of Kish known today as Tell Ingharra (fig.80). Although an inscription on the figure's back is now too faint to decipher, other similar examples with a legible text help to identify him as

Nin-shubur, messenger of the gods who brought news or delivered orders. His insignia of office is the gold staff he carries in his right hand.

Little is known of Babylonian relations with the east where the Medes remained the dominant power. They controlled the region reaching from western Iran through central Anatolia, probably through a network of tribal regions unified under the ruler of Ecbatana as their paramount chief. Some of these Anatolian territories were threatened around 600 BC by Alyattes (610–560 BC) the ruler of Lydia who had extended his authority eastwards. A series of skirmishes and full-scale battles took place between Median and Lydian armies. According to the Greek historian Herodotus, peace was established between the warring parties following a solar eclipse (calculated to have occurred on 28 May, 585 BC) that had been interpreted as a divine message and a bad omen. Herodotus claims that negotiations were led by Nebuchadnezzar and his client ruler of Cilicia and that the treaty was sealed by the marriage of the daughter of Alyattes to a son of the new Median overlord Ishtumegu – known to the Greeks as Astyges.

The solidity of the Babylonian Empire was apparently maintained even with a rapid change of rulers following Nebuchadnezzar's death in 562 BC: his son Amel-Marduk reigned for just two years before he was assassinated by his brother-in-law Neriglissar (559–556 BC). Neriglissar's son Labashi-Marduk, probably just a child, was accepted as king by only a few of the Babylonian cities and he survived for about a month before being killed in a conspiracy of Babylonian nobles. Out of these bloody intrigues Nabonidus (555–539 BC) emerged to claim the throne. Although not related to the royal family, he was an experienced soldier and was probably in his sixties, with a grown son (Bel-shar-usur). The new king faced opposition from some of the major Babylonian cities, but he presented himself as a traditional ruler by installing his daughter at Ur as high priestess of the moon god Sin, a deity to whom he was particularly committed. In Syria Nabonidus had the temple of the moon god rebuilt at Harran, which had lain in ruins since the Medes and Babylonians had wrested it from the Egyptians in 610 BC. It is possible that the king's actions created disquiet in Babylon among the priests of Marduk. However, any opposition to the king, reflected in later anti-Nabonidus propaganda, may have related to his lack of participation in the important cultic rituals in the capital. Having concluded that his presence in Babylon was unnecessary, Nabonidus left his son Bel-shar-usur as ruler in the city. The king meanwhile established himself in northern Arabia for ten years.

Detail of fig.75, Gypsum relief panel showing the deportation of Babylonian prisoners, from the South-West Palace, Nineveh, Iraq, *c.*700–695 BC

# V. From India to Egypt: 500 BC – AD 650

Silver drachma of Shapur I,
*c.*AD 241–270, fig. 103

# V. From India to Egypt: 500 BC – AD 650

AS WE HAVE SEEN, Persian Iranians had probably migrated into eastern Elam, perhaps through the kingdom of Ellipi, by around 700 BC. This mountainous region may have been relatively sheltered from the impact of Assurbanipal's campaigns of 647 BC and, although a weakened Elam had become even more fragmented politically, there had remained a sense of Elamite identity. Indeed, in the vicinity of the unoccupied ruins of Anshan, Elamite heritage seems to have been claimed by the Persian rulers of Parsa (Greek Persis) who gradually extended their authority across the other small kingdoms of south-western Iran. Kurush II, or as he is better known in the west, Cyrus II (559–530 BC) was a member of this dynasty. He described himself as the king of Anshan and his military achievements should probably be understood in the context of the earlier kings of Susa and Anshan and their occasional domination of the Zagros and Babylonia. A good deal of our understanding of the rise of Cyrus is dependent on much later sources, especially the Greek 'Histories' of Herodotus written in the mid-fifth century BC, a hundred years after the events they describe. As a result these sources have to be used with caution. Nevertheless, there are also useful contemporary documents from Babylonia.

Although sharing a common Iranian heritage, the Medes may have felt threatened by the growth of Parsa and around the year 550 BC conflict broke out between the armies of Astyges and Cyrus. According to Babylonian accounts, the Median army rebelled and handed Astyges to the Persians, who then marched to Ecbatana and plundered its riches. By dominating Ecbatana, Cyrus claimed authority over the various Median tribes and, after this initial victory he consolidated his control over their extensive territories by marching through western-central Iran into Anatolia (map 5, right). Here he conquered the powerful and rich Lydian kingdom of Croesus in about 545 BC.

Tensions between the Persians and Babylonians had probably already led to hostilities, perhaps over the region of Susa which once again was united politically with the highlands. As the Medes and Lydia had been allies of Babylonia, it may have been inevitable that battle lines would be drawn between the highlands and lowlands. The Babylonian and Persian armies met at Opis, east of the Tigris, sometime in late September or early October

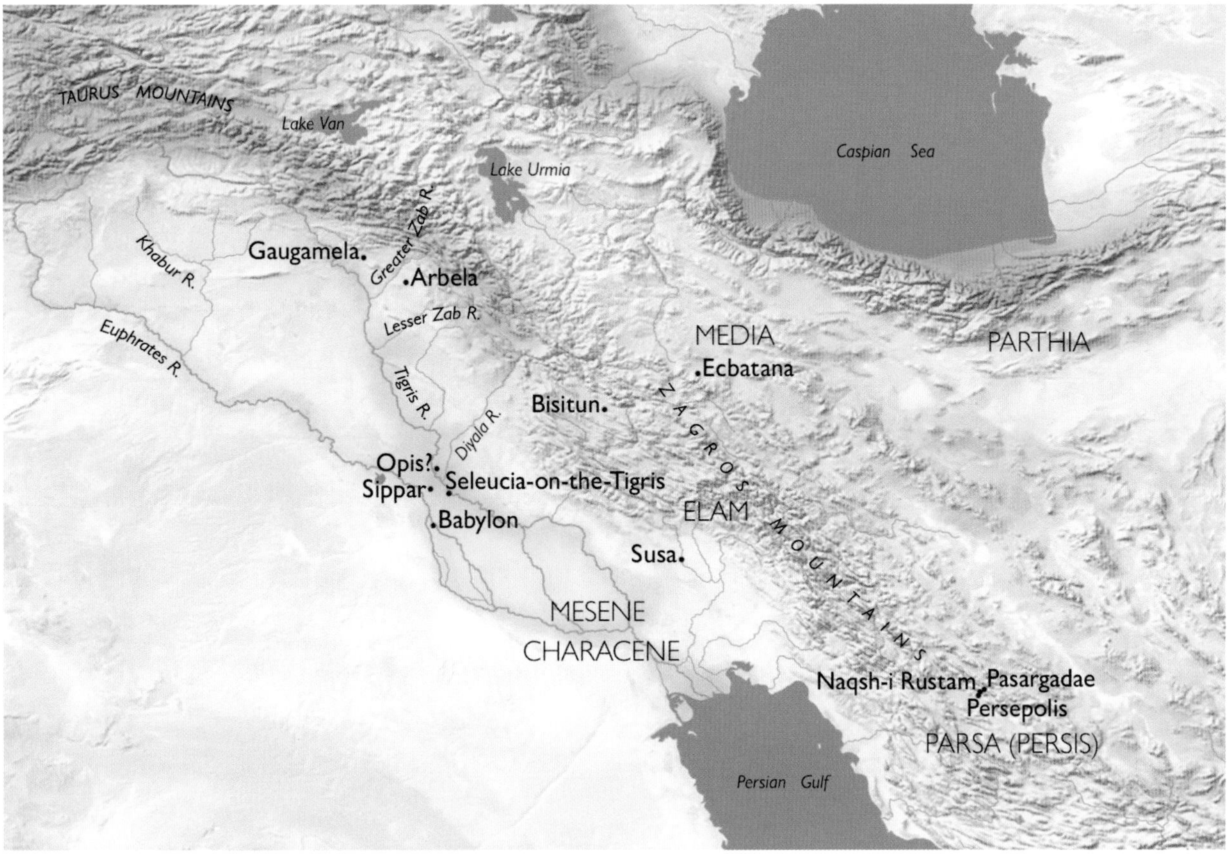


539 BC, along one of the traditional routes taken by earlier Elamite invaders of southern Mesopotamia. Cyrus forced the Babylonian army to withdraw and the Persians plundered Opis, massacring the inhabitants. The city of Sippar surrendered to Cyrus and the Persian general Gobryas led his troops to Babylon, where the gates were opened to them, and Nabonidus, who had returned to the capital, was taken prisoner. Cyrus now made a triumphal entry into the city. In the tradition of earlier Mesopotamian kings he had himself presented as overturning the disasters, imagined or otherwise, of the previous reign and restoring the order demanded by the god Marduk who supported him. This included making generous donations to the temples, the release of people from debt, the return of deportees, and the restitution of divine images. Many of these righteous deeds are recorded in the so-called Cyrus Cylinder from Babylon, a traditional Babylonian cuneiform document, probably composed by the priests of Marduk on behalf of the new ruler (fig.81):

> *He [Marduk] sought and looked through all the lands, searching for a righteous king whose hand he could grasp. He called to rule Cyrus, king of Anshan, and announced his name as king of the universe. He made the Guti-land and all the Medes bow in submission at his feet and so [Cyrus] assiduously looked after the justice and well-being of the Black-Headed People over whom he had been made victorious. And Marduk, the great lord, leader of*

*his people, looked happily at the good deeds and steadfast mind of Cyrus and ordered him to march to his own city Babylon, set him on the road to Babylon, and went alongside him like a friend and companion.*

The Cylinder is written in Akkadian and reveals nothing about Cyrus's own religious views or those of his homeland. Later Persian royal inscriptions portray the kings as sacrificing to the great god Ahuramazda, who required them to re-establish order on earth and defeat the 'lie'. It is likely that Cyrus also followed these beliefs which may have been based on an oral religious tradition that had taken shape among the Iranians in Central Asia during the second millennium BC – this was to become the Avesta, the sacred texts of Zoroastrianism (named for Zoroaster, the Greek name for the prophet Zarathustra) that were probably written down towards the end of the Sasanian period in the sixth–seventh centuries AD.

**FIG. 81** Baked clay 'Cyrus Cylinder' from Babylon, Iraq, *c.*539 BC, l. 21.9 cm, 1880,0617.1941 © Trustees of the British Museum

FIG. 82 Tomb of Cyrus II at Pasargadae, Iran, c.530 BC, ht. 11 m © Roger Wood/Corbis

The importance of Babylon for the Persian ruler, symbolically as well as politically, was emphasised when Cyrus appointed his son Cambyses as king in 538–537 BC. However, after a year the arrangement was abandoned and the territories of Babylonia became a single vast province under a Persian governor. Indeed it was not in Mesopotamia but in the heartland of Persia that Cyrus built his own royal centre. His conquests had brought the ruler access not only to great wealth but also to skilled workers, and Cyrus may have transported to Persia local materials and artists from Ionia and Lydia in western Anatolia, possibly including stonemasons working on the unfinished temple of Artemis at Ephesus. They were set to work to build Pasargadae located around 40 kilometres north-east of the ruins of Anshan. The city was laid out with palaces containing columned halls and porticoes, surrounded by gardens with stone-lined water channels. Stone door jambs and other elements were carved with images of guardian figures which drew on Elamite, Egyptian, and Assyrian prototypes. Many of the sculptures have cuneiform inscriptions, representing some of the earliest written texts in any Iranian language and known today as Old Persian; although these name Cyrus as the builder, they were almost certainly added during the reign of Darius I.

Cyrus now turned his attention to the extensive eastern borders of his realm that perhaps reached as far as modern Afghanistan and Turkmenistan. Here, the later Greek authors claim he was killed on campaign. His body was brought back to Pasargadae and interred in a magnificent stone tomb comprising a single chamber set on a stepped platform; the entire structure once stood over eleven metres high (fig.82).

Cyrus was succeeded by his son Cambyses (529–522 BC) who devoted the early years of his reign to preparations for containing the expansionist aspirations of Egypt. A Persian navy and land army pushed to the Egyptian capital of Memphis where a garrison was installed and from where troops moved south to ensure control of the entire country. However, in 522 BC a rebellion in Parsa may have forced Cambyses to return to the east but he died en route. It is possible that his brother Bardiya had tried to seize the throne but was killed in the attempt. The true nature of the two men's deaths is unclear, since the details are shrouded in an elaborate story which was later disseminated throughout the empire, perhaps to cover the tracks of the rebels. This account asserts that just before Cambyses died he admitted that he had secretly arranged for Bardiya to be killed. Taking advantage of the situation, an impostor called Gaumata, who bore a striking physical resemblance to Bardiya, declared that he was Cambyses' brother and claimed the Persian throne. This deception was recognised by a Persian noble, Darius, who killed Gaumata and, having demonstrated his worth by recognising the 'lie', became king. In reality it is possible that Darius had conspired against both Cambyses and Bardiya.

## THE ACHAEMENID PERSIAN EMPIRE

The new ruler, Darius I (521–486 BC) – who was probably not a member of the royal family but nevertheless claimed a connection through a common ancestor named Achaemenes, hence the Achaemenid Persian Empire – was himself now faced with a series of substantial rebellions that threatened to fragment the empire. Significantly, among the places that witnessed uprisings were Elam, Babylonia, Media, and even Parsa itself – perhaps the clearest indication that Darius was perceived to be a regicide and usurper. With support from among the Persian nobles, Darius managed to crush all opposition but it took him over a year to do so. He commemorated this great triumph with a huge relief carved on a cliff face at Bisitun, overlooking the Great Khorasan Road (fig.83). The imagery was probably inspired by the Sar-e Pol-e Zahab relief (see fig.51) a little further along the road, since the Persian king had himself depicted in the traditional heroic pose, crushing Gaumata under his foot, with representatives of nine rebellious peoples roped together at the neck before him. Above the rebels is a figure in a winged disc, representing the great god Ahuramazda or perhaps royal or divine 'glory'. A cuneiform inscription with an account of Darius's rise to power and his military successes was added to the scene. This was written

in Elamite, highlighting the royal connections to this ancient kingdom. At a later point in his reign Akkadian and Old Persian versions were also inscribed on the cliff. The latter was written in a cuneiform script developed under Darius and disseminated through the empire, probably to strengthen its Persian character.

FIG. 83 Rock relief of Darius I at Bisitun, Iran, *c.*520 BC, W. 5.5 m


Although it had been necessary to crush several rebellions in the region of Elam, Darius chose the ancient city of Susa as a royal centre. Here a mudbrick building of monumental proportions was constructed, modelled on earlier Assyrian and Babylonian palaces and combined with a Persian style *apadana* (columned hall). The palace was decorated with elaborate

**FIG. 84** Glazed tiles from the palace of Darius I at Susa, Iran, *c.*500 BC, ht. 220 cm, AOD487 © RMN-Grand Palais (musée du Louvre) / Franck Raux

panels and scenes made from plain and glazed bricks, including rows of lance-bearing guards – perhaps the king's personal bodyguard of 10,000 lancers, the so-called 'Immortals' (fig.84). On large clay tablets buried within the foundations of the building, Darius recorded in Old Persian how he brought materials and workmen from all parts of the Persian Empire to create the building:

> *The stone-cutters who wrought the stone, those were Ionians and Sardians. The goldsmiths who wrought the gold, those were Medes and Egyptians. The men who wrought the wood, those were Sardians and Egyptians. The men who wrought the baked brick, those were Babylonians. The men who adorned the wall, those were Medes and Egyptians.*

In Persia itself, around 40 kilometres to the south-west of Pasargadae in the Marvdasht Plain, Darius founded another centre; this was to be the dynastic capital that would be elaborated further by several of his successors.

It was called Parsa; the Greeks named it Persepolis. The buildings were arranged on a raised stone terrace measuring about 455 by 300 metres. Enormous structures erected there included the Apadana, with its main hall containing stone columns twenty metres in height and, nearby, the so-called Hall of 100 Columns that would have been able to accommodate thousands of people (fig.85). Stone sculptures and reliefs decorated many of the doorways and staircases, depicting the enthroned king with Ahuramazda hovering above, a royal hero figure in combat with monsters and bulls, and rows of nobles in Persian and Median dress, guards, and attendants. The small stone fragment illustrated here depicts the head of one such bearded Persian (fig.86). Other reliefs included people from twenty-three regions, designed to express the extent of the empire, bringing presents and tribute or supporting the king's throne. This notion of an empire of diverse people united under the Persian king was a central feature of royal ideology. All of the reliefs were probably originally painted and some were decorated with ornaments such as bracelets and crowns of gold. As at Pasargadae, much of this sculpture

**FIG. 85** View of the east stairway of the Apadana at Persepolis, Iran © Gérard Degorge/Corbis

**FIG. 86** Fragment of a limestone relief depicting a Persian, from Persepolis, Iran, *c.*500–450 BC, W. 15.3 cm, AN1982.944 © Ashmolean Museum, University of Oxford

was created by Greek and Lydian craftsmen although among the artistic influences at Persepolis were the long abandoned monuments of the Assyrian imperial palaces, especially the colossal gate guardians that must still have been visible at Khorsabad or Nineveh. It would be close to Persepolis that the tombs of the Achaemenid kings would be cut into vertical rock faces. Seven are known, the most famous being those at Naqsh-i Rustam (fig.87).

The immense imperial territories – covering around 7,500,000 square kilometres from Libya and Thrace to Central Asia and India – were divided into provinces, referred to by the Iranian term 'protector of the kingdom' and adapted into Greek as 'satrapies'. Each province was governed by a satrap (governor): a Persian or Iranian noble generally held the highest posts but close links were developed with the local elite. The administrative and military centre of each satrapy was often the same as the old capital of the conquered territories. Thus in Media, the capital was Ecbatana where, according to Greek authors, the Achaemenid kings had a summer capital and treasury, while in Mesopotamia, the capital remained at Babylon. Probably

FIG. 87 View of the Achaemenid royal tombs at Naqsh-i Rustam, Iran © Ggia (Own work) [CC BY-SA 3.0 (http://creativecommons.org/licenses/by-sa/3.0)], via Wikimedia Commons

early in the reign of Xerxes (485–465 BC), the lands of the former Babylonian Empire were divided into two new satrapies: Beyond the River (west of the Euphrates reaching to the Egyptian border) and Babylonia (the whole of Mesopotamia). Assyria may have been something of a rural backwater under the Achaemenids, with very limited occupation at the old Assyrian sites, although it is possible that Arbela may have been an important hub. Each capital contained a palace used by the satrap himself but it was also maintained for use by the king as he progressed across the empire. Nebuchadnezzar II's old palace in Babylon was used as the seat of the Persian satrap although, late in the fifth century BC, an elegant columned palace (*apadana*) was constructed nearby. The Persian kings followed local traditions and ideology so that in Babylon they maintained and rebuilt temples, and performed the appropriate rituals, especially the important New Year Festival.

To facilitate communications across the empire, the network of dirt roads with stopping places that had already been established during the Assyrian Empire was expanded. The principal artery was the so-called Great Royal Road, the long-established highway that linked Susa along the piedmont to the Diyala/Tigris confluence (and its branches to Babylon and Ecbatana), then north to Arbela and, passing the remains of the Assyrian capitals, headed either across the Jezirah and Syrian steppe or north into Anatolia where the road ended at Sardis. Way stations were built at regular intervals and were provided with sufficient stock for issuing rations to travellers on official missions. Although more formalised, the route would have varied depending on the time of year and the weather, especially when this affected the water levels in the rivers, altering the accessibility of crossing points.

Judging from the surviving documents, certain segments of Babylonian society seem to have profited enormously under the empire. Babylon itself became a vast cosmopolitan city with people from Afghanistan together with Arabs, Armenians, Carians, Egyptians, Greeks, Indians, Iranians, Jews, Lydians, Phrygians, and Scythians. Wealth was generated by large-scale irrigation projects that the Persian rulers promoted across the Babylonian plains, but also in northern and south-western Iran. The area of Parsa itself, which had been a sparsely settled region of cattle herders, was transformed by the end of the empire into a densely settled, fertile area. Within the Zagros, the populations were probably never fully integrated into the imperial structure. Local leaders were placed under obligation to the Persian king who presented them with gifts in return for manpower and the security of routes through the mountains.

Without accounts comparable to the Assyrian royal inscriptions and correspondence, much of our knowledge about the political history of the Achaemenid Empire derives from Greek sources that inevitably focus on the western territories and especially on the Persian defeats at Marathon in 490 BC and Salamis and Platea in 480–479 BC. While it had proved impossible to extend Persian control further in Greece, much more serious were the revolts that had broken out in Babylonia in 481 and 479 BC, perhaps connected with Xerxes's restructuring of the satrapy. The loss of the province would have been catastrophic to the empire and the Persian king moved rapidly to crush opposition. This was successful and there are no other accounts of rebellion within Babylonia. Babylon itself continued to flourish – the later stories that Xerxes punished the city by destroying the temples and removing the statue of Marduk are certainly apocryphal.

Although the family of Darius I would never be effectively challenged as rulers of the empire for over 200 years, there was the recurring problem of court conspiracies, which the Greek sources are only too happy to recount, often in lurid detail. In August 465 BC, both Xerxes and his crown-prince were murdered. This was almost certainly engineered by one of the king's sons who took the throne as Artaxerxes I (465–424 BC). After a long reign, he was succeeded by a bastard son, Darius II (423–405 BC), but his claim to the throne was disputed violently by two legitimate sons. Ultimately, Darius was successful against his rivals and they were killed. Of events outside the royal court during his reign, the Greek accounts speak of a series of wars against a tribal group called the Cadusians. They dominated an area in northern Iran, south-west of the Caspian Sea and this state of affairs would be a recurring problem for the Persians into the mid-fourth century BC.

The best known of the internal court revolts followed the accession of Darius's son to the throne as Artaxerxes II (404–359 BC). He was challenged by his brother Cyrus who had gained some support from within Persia but relied on a mercenary army of some 10,000 Greeks. Among the soldiers was the Athenian Xenophon who wrote a lively account of the expedition known as *Anabasis* ('An ascent'). He describes how the Greek army made its arduous way from western Anatolia into Mesopotamia and then followed the Euphrates into northern Babylonia. Here in 401 BC it met Artaxerxes's forces at Cunaxa, where it was rapidly defeated and Cyrus was killed. Stranded in Persian territory, the remnants of the army struggled northwards along the east Tigris route to Anatolia and the Black Sea.

FIG. 88 Gold armlet from the 'Oxus Treasure' from Takht-i Kuwad, Tadjikistan, c.500–350 BC, ht. 12.3 cm, 1897,1231.116 © Trustees of the British Museum

Following his success against Cyrus, Artaxerxes now faced one of the most momentous challenges in the history of the empire – at least of those we know about – when in 400 to 399 BC Egypt rebelled and seceded. The country would not be regained by the Persians until 343 BC when, after a long and gruelling campaign, Artaxerxes III (358–338 BC) captured the rebellious city of Sidon on the Mediterranean coast and then reconquered Egypt. The empire was thus reunified in all its diversity, as not only celebrated in the reliefs at Persepolis but also recognised in the use of multilingual inscriptions on royal objects and monuments. At a more basic level, a sense of unity was expressed through the use of Aramaic as the written language of administration throughout the empire. At the level of the court, however, prestigious objects created a shared identity. While a wide variety of styles and forms existed there was also a recognisably Achaemenid court style, probably promoted outside Iran by satraps and other senior administrators. Many of the most elaborate objects are represented on the reliefs at Persepolis being brought as tribute. Among them are extraordinary armlets such as this one belonging to the so-called Oxus Treasure that originates from the region of Takht-i Kuwad in modern Tadjikistan (fig.88). The hollow spaces would have contained inlays of glass or semi-precious stones. Xenophon mentions that armlets were among the items considered as gifts of honour at the Persian court. Other rich objects found throughout the empire were silver vessels such as this shallow bowl with gadrooned lobes, a type, often described by the Greek term *phiale*, which was used for drinking across the Near East (fig.89). Forms of precious metal vessels were also produced in terracotta versions such as the horn-shaped *rhyton* terminating in an animal's head, which would have been used both as a drinking cup and as a pourer for wine (fig.90).

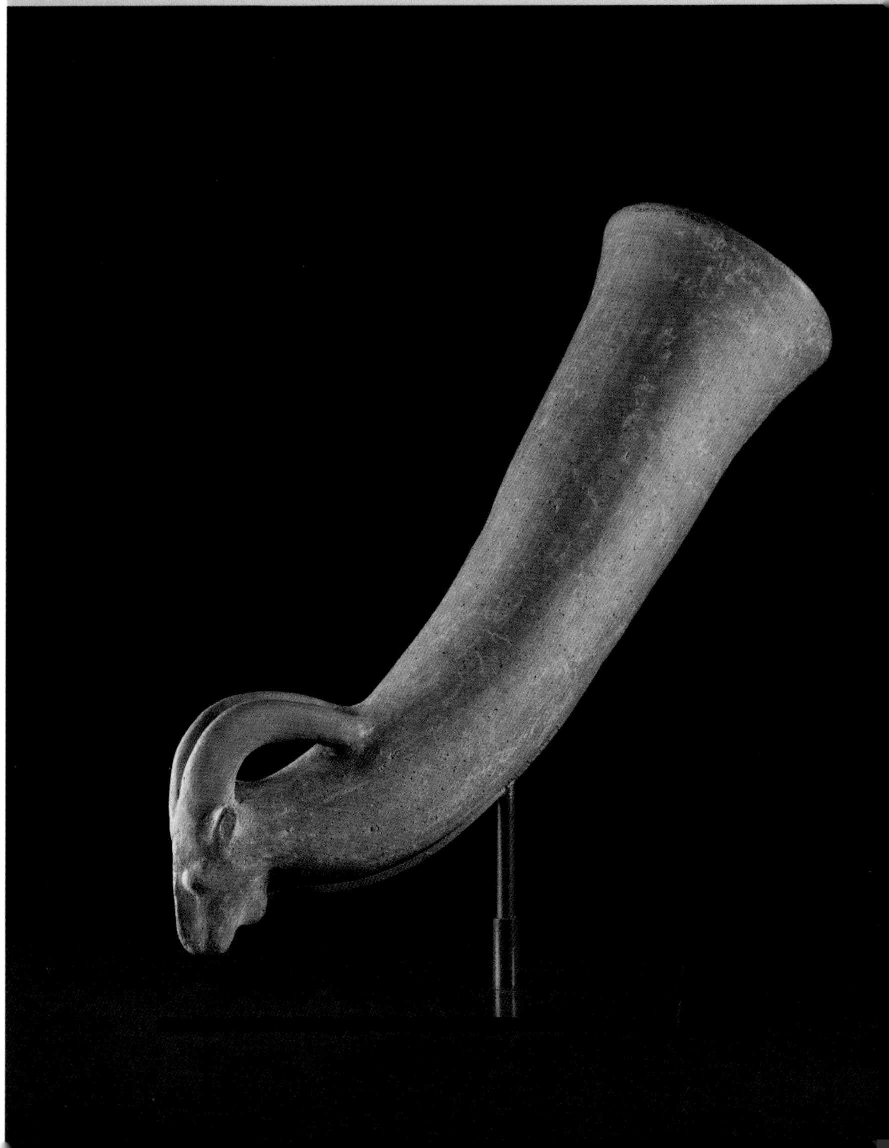

TOP:

**FIG. 89** Silver phiale, *c.*500–350 BC, diam. 17.2 cm, AN1971.988 © Ashmolean Museum, University of Oxford

BOTTOM:

**FIG. 90** Terracotta rhyton, from Deve Huyuk, Syria, *c.*500–400 BC, l. 34.6 cm, AN1913.636 © Ashmolean Museum, University of Oxford

**FIG. 91** Bronze Boeotian helmet from near Sert, Turkey, *c.*400–200 BC, ht. 24 cm, AN1977.256 © Ashmolean Museum, University of Oxford

Later Greek sources would have us believe that the final years of Achaemenid kingship were mired in yet more bloody intrigues. According to these accounts, Artaxerxes and his family were murdered in 338 BC during a palace coup orchestrated by a court eunuch called Bagoas. The one surviving royal prince, still only a boy, was placed on the throne as Artaxerxes IV. After only two years, however, Bagoas had him murdered and replaced by a more distant member of the royal family, Artashata. The new king took the throne name of Darius III (335–331 BC) and, once he had consolidated his position, had Bagoas killed. It would be Darius's challenge to confront the invading army of Alexander III of Macedon. Despite the rapid change of rulers, there is nothing to suggest that the empire had been weakened or divided. In fact, its solidity would be demonstrated by the resistance faced by Alexander, requiring him to take nearly every satrapy by force.

## THE MACEDONIAN INVASION AND CONQUEST

In the spring of 334 BC Alexander sailed an army across the narrow strait of the Hellespont that divided Macedonian territories from Persian Anatolia and, as later reports would have it, he hurled a spear onto the beach of Troy. With this symbolic act, recalling Greek victories in the Trojan War and avenging Xerxes's invasion of Greece in 480 BC, began a decade-long invasion and takeover of the Achaemenid Empire. Many of the details of the campaign are known only through much later literary and biased accounts. It is clear, nonetheless, that the fast moving Greco-Macedonian infantry phalanx, wielding long pikes (*sarissa*), together with heavy cavalry – wearing so-called Boeotian helmets such as the one illustrated here that was pulled from the River Tigris in south-eastern Turkey in the nineteenth century (fig.91) – proved highly effective against the massed, slower Persian forces;

under Alexander's leadership they proved decisive. This was demonstrated as early as the beginning of April when a Persian army was defeated at the Granicus River close to the ancient site of Troy. Having secured the western coast, Alexander crossed through Anatolia. In November 333 BC another large Persian army, this time led by Darius, was defeated at Issus close to an important route through the Amanus Mountains into Syria. Darius escaped and the Macedonian army marched south, besieging the island city of Tyre and massacring its male inhabitants, before moving on to Egypt where Alexander was welcomed as a liberator in 332 BC. The following year he headed back through Syria into northern Mesopotamia. Here, in October 331 BC, after crossing the Tigris, Alexander's forces engaged an immense Persian army close to the site of Nineveh at Gaugamela. A Babylonian astronomical diary records the events of that month:

> *The twenty-fourth [day], in the morning, the king of the world [Alexander] [erected his] standard. Opposite each other they fought and a heavy defeat of the troops [of the king he inflicted]. The king [Darius], his troops deserted him and to their cities [they went]. They fled to the land of the Guti.*

Darius retreated via Arbela through the mountains to Ecbatana.

Alexander now marched on Babylon where the Persian satrap Mazaeus peacefully handed the city over to him – in so doing he managed to retain his position, albeit under the watch of a Macedonian garrison and officers. As in Egypt, Alexander presented himself as a model king who restored local customs and traditions including paying for a refurbishment of the temple of Marduk and performing appropriate sacrifices. In pursuit of Darius, Alexander left Babylon for Susa where the satrap of Elam, encouraged perhaps by the king's lenient attitude in Babylon, also opened the city's gates to him. An impressive treasure of something approaching 49,000 talents (about 1,470 tons) of silver was seized here. Heading east into the Zagros, Alexander entered the territory of the Uxian tribe. They demanded payment for passage just as they always had of the Persian kings. However, the Macedonians outmanoeuvred them, taking their main settlement and slaughtering the inhabitants. By January 330 BC the Macedonian army had reached the so-called Persian Gate (modern Tang-e Meyran), a narrow valley that gave access through the Zagros to Persepolis. This they found blocked by an army under the satrap of Parsa, Ariobarzanes. The Persians held out for a month but Alexander, after considerable losses, eventually encircled them, broke through the defences and overran them. The defeat of Ariobarzanes's forces removed the last obstacle on the way to Persepolis. Upon his arrival, Alexander seized the abundant contents of the city's treasury; this, together with the wealth from Susa, meant that he no longer needed money from the west to maintain the campaign. According to later Roman accounts, Persepolis was turned over to his soldiers to plunder. After a four-month sojourn in the city, many of the palaces and audience halls on the terrace were set alight and destroyed. The ancient sources do not agree about why this took place and explanations range from a deliberate act of revenge for the earlier Persian burning of the Acropolis of Athens, or an impulsive, drunken act, to an expression of Alexander's anger at not being recognised as the legitimate successor to Darius. Since the Persian king was still alive, the ruin of his capital would have certainly denied him an ideological centre, but, as Alexander would come to realise, it also damaged his own claim to rule the empire. Ironically, the destruction layer at Persepolis is the only archaeological evidence of Alexander's presence in Persia.

By July 330 BC, the Macedonians had closed in on Darius who was moving eastwards, probably along the Great Khorasan Road south of the Caspian Sea. It was at this point that the Persian king was betrayed by one of his most senior nobles, Bessus the satrap of Bactria (northern Afghanistan). Bound and bundled into a waggon, Darius was stabbed to death as the Macedonian forces approached. Withdrawing to Bactria, Bessus proclaimed himself the new Persian king (as Artaxerxes V), and thereby forced Alexander to follow him in order to suppress threatened rebellions in the eastern satrapies.

By 326 BC, after campaigns in Central Asia and the Indus Valley, Alexander turned back to the west. The overland journey along the southern coast of Iran was long and difficult but, by the beginning of 324 BC, the decimated army had reached Pasargadae; Alexander is said to have visited the tomb of Cyrus (see fig.82) and discovered that it had been damaged and robbed. On returning to Susa, via Persopolis, Alexander accused many of the Persian nobles he had left in power of disloyalty and there followed a bloody purge – Greeks and Macedonians were appointed in their place. Alexander now adopted the role of the Achaemenid ruler. He repaired the tomb of Cyrus, expressed regret at the burning of Persepolis, and conducted a mass wedding of Macedonian officers to noble Iranian women in the Persian manner. When he attempted to send veterans home there was a mutiny of soldiers at Opis who thought they were being abandoned by the king in favour of the Persians. From Susa Alexander progressed to Ecbatana and then, in early 323 BC, he set out for Babylon – on the way through the mountain passes his forces confronted and defeated the Kossaean people.

In Babylon Alexander died of unknown causes in early June 323 BC after suffering a high fever for ten days.

Alexander left behind an impressive empire in which he had only just begun to build a sense of cohesion. Some Achaemenid structures and ceremonies were maintained but he introduced others which were to have long lasting impact: new cities were founded with mixed Greek and local populations (though not in Iran); there was a fusion of Greek and local elites established through marriage and among at least some of his satraps an attempt at integration. Peucestas, the satrap of Parsa, had adopted Persian clothing and learnt the Persian language. Near Eastern armies were also trained according to Macedonian principles. Arguably, among the most important novelties Alexander introduced was coinage. Although coins had been minted and circulated in the western half of the Persian Empire, their use was often limited or non-existent. Alexander's coins were issued in his name in all regions of the empire. One of the mints was established at Babylon where this silver tetradrachm was struck during the king's lifetime (fig.92). It depicts on the obverse the head of a young Herakles, and on the reverse a seated Zeus alongside Alexander's name in Greek. The transformation of the Persian bullion into coinage must have had a profound impact on the local economies. The widely circulated coins established a cultural uniformity through their images and associated Greek inscriptions. Despite these unifying elements, however, Alexander's death without an adult heir left the empire prey to a violent generation-long power struggle known as the Diadochi ('Successor') Wars.

**FIG. 92** Silver tetradrachm of Alexander III of Macedon struck at Babylon, *c*.330–323 BC, diam. 2.6 cm, HCR3059 © Ashmolean Museum, University of Oxford

Initially Perdiccas was appointed from among Alexander's generals as regent and the satrapies of the empire were divided between the rest. By 321 BC, however, rivalry among the new satraps had led to a series of wars, the assassination of Perdiccas and a new division of the empire: Seleucus was appointed as satrap of Babylon under the new regent Antipater. Nonetheless, within only a few years the wars between the satraps had resumed: Peithon, satrap of Media, attempted to extend his authority over eastern Iran but was opposed by a coalition of Iranian and Central Asian governors. The most serious confrontation was between the armies of the generals Antigonus and Eumenes, fought in 316 BC in Susiana and Media. Seleucus fled before the victorious forces of Antigonus who was left as undisputed ruler of the whole of Western Asia.

## THE SELEUCID EMPIRE

In 311 BC Seleucus returned to Babylonia at the head of an army and, with local support, seized back the satrapy, as well as those of Susiana, Media, and Parsa. Antigonus responded with a series of campaigns but was unsuccessful in retaking the territory. Between 308 and 304 BC Seleucus expanded his rule across the eastern satrapies of Iran and Central Asia through conquest and diplomacy. His marriage to Apame, an Iranian noble-woman from Sogdia, probably helped to win him support from the local elite, and while in India he eventually established an alliance with the Mauryan emperor Chandragupta. With the defeat of Antigonus at the battle of Ipsus in 301 BC, Seleucid territories extended into northern Syria and Anatolia. Alexander's empire had thus become effectively divided between Western Asia under Seleucus I Nicator ('the Victorious') and the Egyptian and North African territories of Ptolemy I Soter ('the Saviour').

The impact of these battles on the local populations of Mesopotamia and Iran must have been profound. Yet there is no record of any indigenous anti-Macedonian revolts. This may have been partly the result of the successful cultivation of local support, and policies of integration pursued by the satraps. Caution is necessary however as inevitably our evidence for reconstructing the history of the Seleucid Empire comes largely from Greco-Roman writers who only rarely make reference to events beyond Anatolia and the eastern Mediterranean. Nonetheless, the imperial centre of power at this time certainly shifted from Persia to Babylon and westwards to the newly founded cities of Seleucia and Antioch in Syria.

Babylon lost its position as the preeminent city in Mesopotamia when the city of Seleucia-on-the-Tigris was founded, probably before 301 BC.

Approximately 60 kilometres to the north of Babylon, at the confluence of the Tigris and a major canal connecting it with the Euphrates, Seleucia was described as the 'city of kingship' and became the central hub of the empire. Nevertheless, Babylon remained an important regional city, where a garrison and officials were stationed. Susa too continued to be occupied and was re-founded as Seleucia-by-the-Eulaios – Elam and the Elamites continue to be recognised in Greek and Latin texts in the form of Elymais and Elymaeans. Like Susa, which was probably established as a military colony, other garrison cities of Greeks and Macedonians were now founded in Iran. Many were located along the Great Khorasan Road linking Seleucia-on-the-Tigris to Ecbatana and then from Media to Central Asia, as well as on the routes to the head of the Persian Gulf and its maritime trade.

The period of Seleucid rule saw aspects of Greek culture being gradually adopted – it was not imposed as a deliberate policy. Indeed, there appears to have been a close relationship between the Seleucid administrators and the local elite with no apparent opposition between native culture and Greek culture: individuals were familiar with both and moved easily between them depending on the context, within a remarkably fluid world of beliefs and practices. This is evident in the use of terracotta female figurines. They were made in moulds, widely accessible and cheap to replace. In the Greek world the figurines were used as votive objects in temples and were generally depicted clothed. In Babylonia, clay figurines of females had traditionally been shown frontally and nude and were probably connected with personal religion in the home. Under the Seleucids Greek-style figurines were probably not used as votives and many examples have been found in graves. While some are shown with Greek hairstyles, others wear a conical, Near Eastern headdress, such as this example from Kish where the figure is also nude to below the waist (fig.93).

When Seleucus I was assassinated in 281 BC revolts broke out in Anatolia and Syria but Mesopotamia and Iran appear to have remained peaceful under his son and successor Antiochus I (281–261 BC) – he was, after all, half Persian through his mother Apame and had established close ties with the Iranian nobility. Antiochus consolidated Seleucid rule and much of his reign was devoted to long and violent wars with the Ptolemies of Egypt over control of the east Mediterranean seaports. This conflict continued into the reign of Antiochus II (261–246 BC) and, taking advantage of the king's western focus, his satrap of Parthia, to the south-east of the Caspian Sea, gradually seceded from Seleucid control. According to the Latin author Justin (writing sometime between the second and fourth century AD), the satrap of Bactria also broke free. From the middle of the third century BC, the empire

FIG. 93 Terracotta figurine of a reclining woman, from Kish, Iraq, *c.*300–200 BC, l. 9.6 cm, AN1924.239 © Ashmolean Museum, University of Oxford

gradually took the form of a number of increasingly independent vassal states around an imperial centre consisting of Mesopotamia and Syria. Relationships between the core and the periphery were cemented by gifts and marriages.

In Parthia, again according to Justin:

> *One Arsaces, a man of uncertain origin, but of undisputed bravery, happened to arise at this time and ... took upon himself the government of the country.*

Arsaces I had gained control of the satrapy around 247 BC with the help of the Parni tribe (who were probably nomadic and had moved into northern Iran from inner Asia). The Roman sources call his kingdom – and the empire that would grow from it – 'Parthian' after the region of its origin. This name might suggest erroneously a specific ethnic group and many modern studies prefer 'Arsacid' as the name for the empire and its dynasty. The successors to Arsaces ruled from a capital at Nisa (in modern Turkmenistan, close to the border with Iran) over a sedentary, urbanised society with wide-ranging contacts, not least along the part of the Great Khorasan Road that they controlled south of the Caspian Sea. The chronology of the rulers is less well understood than that of their contemporaries to the west because there are so few surviving Arsacid written sources, leaving us heavily reliant on later Latin authors – some useful information can be gathered from the production and circulation of individual rulers' coinage.

In Mesopotamia, the Seleucid rulers maintained their attention to the western half of their empire, which was seriously threatened under Seleucus II (246–225 BC) when Ptolemy III's troops advanced as far as Babylonia before being forced to retreat. There were also recurrent battles in Anatolia between the army of Seleucus and that of his brother. The empire then faced a revolt in the east during the short reign of Seleucus III (225–223 BC), when Molon, satrap of Media, and his brother Alexander of Parsa claimed independence, expanding their control into Susiana and Mesopotamia as far as the Euphrates River in Syria. The new Seleucid ruler Antiochus III (223–187 BC) – here depicted in profile wearing a diadem on a silver tetradrachm, with an image of the Greek god Apollo on the reverse (fig.94) – therefore faced a formidable challenge. He crushed the Molon rebellion and turned to confront Ptolemy IV in Syria but here his armies were defeated.

Returning to the east around 211 BC Antiochus embarked on a long yet successful campaign through the satrapies of northern Iran, Bactria, and Gandara (eastern Afghanistan).The Arsacid king Arsaces II was defeated in battle but was subsequently recognised as a vassal king. By 202 BC Antiochus had returned to the west where war had resumed against the Ptolemies of Egypt. He invaded Greece in 192 BC but was forced to withdraw by Rome

**FIG. 94** Silver tetradrachm of Antiochus III, possibly struck at Edessa, Turkey, *c.*223–187 BC, diam. 2.9 cm, HCR9393 © Ashmolean Museum, University of Oxford

and its allies. Two years later Antiochus was decisively defeated by the Romans at the Battle of Magnesia in western Anatolia. Under the terms of the Treaty of Apamea (188 BC) he was forced to give up all territory north of the Taurus Mountains and pay a considerable indemnity to Rome. The desperate financial situation probably drove him to plunder a wealthy temple in the mountains of Elymais (Elam) in 187 BC but he was killed while fighting the Elamites who may have begun a push for independence.

Although the Seleucid Empire was weakened it was far from being at the point of collapse. Antiochus IV (175–164 BC) was able to advance his armies into Egypt itself even though he was forced to then withdraw when threatened by Rome. He was also faced by a rebellion of orthodox Jews (the Maccabean Revolt) in Judea. Antiochus campaigned successfully against Armenia but, like his father, financial needs propelled him to loot the temple in Jerusalem but when he tried to do the same in Elymais he was driven back. His death in 164 BC cut short any plans he may have had to tackle an emerging threat in the east: the rise of the Arsacids. Under Mithridates I (165–132 BC) they took over from the Ptolemies as the Seleucid's principal military opponents.

Over the following decades the former satraps of Media, Elymais, and Mesene (southern Mesopotamia) broke free from Seleucid authority – Kamniskires, the ruler of Elymais, even plundered Babylonian cities in 145 and 141 BC. These kingdoms straddling the Zagros shared common cultural features even if politically and linguistically (Aramaic in the lowlands and Iranian or perhaps Elamite in the highlands) they were divided. The symbolism of Herakles was especially popular as represented by this figurine that depicts the standing demigod, nude apart from a lion's skin draped over his left shoulder (fig.95). In his right hand he holds a club resting on the ground and in his left hand what may be an apple from the Garden of the Hesperides, referring to one of Herakles's twelve labours. (The figurine was purchased in Baghdad but was said by the seller to have been found at Uruk.

LEFT:

**FIG. 95** Painted terracotta statuette of Herakles, *c.*300–100 BC, ht. 18.1 cm, AN1923.303 © Ashmolean Museum, University of Oxford

RIGHT:

**FIG. 96** Sculpture of reclining Herakles at Bisitun, Iran, 148 BC, ht. 190 cm © Tibor Bognar/Corbis

This is certainly possible as the city was flourishing in the Seleucid period.) A more dramatic representation of Herakles is a stone sculpture overlooking the Great Khorasan Road at Bisitun (fig.96). Dated to 148 BC it is the only known example of monumental Seleucid art in Iran. The accompanying Greek inscription reads:

> *In year 164 [of the Seleucid era], in the month of Panemos, Hyakinthos son of Pantauchos dedicated this statue of Herakles Kallinikos ['Glorious in victory'] for the safety of Kleomenes, the governor of the Upper Satrapies.*

## THE ARSACID EMPIRE

It was against such kingdoms that Mithridates moved in the last decade of his reign, conquering Media and occupying Ecbatana in 147 BC. In 141 BC the Arsacid ruler led his armies through the Zagros passes onto the Mesopotamian plains and entered Seleucia-on-the-Tigris. He withdrew to quell a nomadic incursion in the east but returned to capture the Seleucid ruler Demetrius II (146–140 BC) and the city of Susa (by 138 BC). In about 130 BC the Seleucids under Antiochus VII (138–129 BC) counterattacked and recaptured Babylonia and Media but the king was defeated and killed

**FIG. 97** Silver tetradrachm of Mithridates II struck at Seleucia-on-the-Tigris, *c.*123–88 BC, diam. 3.1 cm, HCR7059 © Ashmolean Museum, University of Oxford

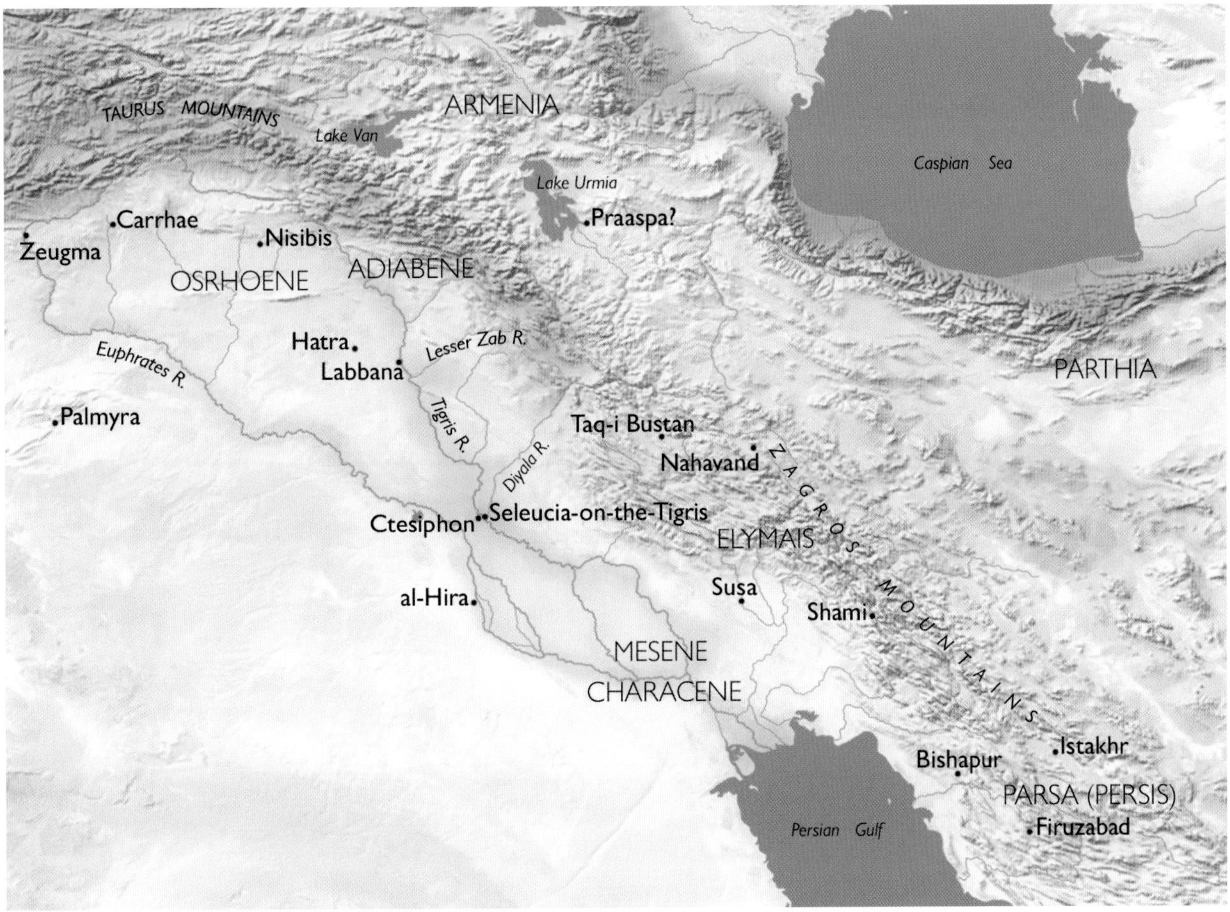


by an Arsacid army under Phraates II (132–126 BC) the following year. Nevertheless, this remained a troubled period for the Arsacid ruler with great pressure on the eastern borders of the empire from nomadic Saka (Scythian) groups, continued resistance from the independent kings of Elymais, and a revolt in southern Babylonia (the kingdom of Mesene-Characene) led by Hyspaosines (map 6, above).

It was Mithridates II (123–88 BC) who would stamp his authority over these regions and then go on to extend the borders of the Arsacid Empire in all directions. The bearded king, wearing a diadem, is depicted on this silver tetradrachm minted at Seleucia-on-the-Tigris; an archer on the reverse holds a bow surrounded by a square four-line Greek inscription (fig.97). By 120 BC Arsacid control of Babylonia had been re-established and Mithridates was able to concentrate on protecting the empire's eastern borders in Central Asia against the Saka and dominate the strategic eastern extension of the Great Khorasan Road leading through Merv, Bukhara, and Samarkand. Indeed, around 115 BC an embassy from the Chinese Han emperor established relations with the Arsacid king and laid the foundations for the so-called 'Silk Road'.

To the west of the Zagros, the small kingdoms of northern Mesopotamia and Syria were recognised as Arsacid vassals – Adiabene (the region of ancient Assyria between the Greater and Lesser Zab rivers) and Osrhoene (between the Khabur and Euphrates rivers). Having defeated the king of Armenia (the region of ancient Urartu) around 105 BC, Mithridates established Prince Tigranes on the Armenian throne. The permanent loss of Iran and Mesopotamia effectively ended the Seleucid state and the Arsacid Empire's centre shifted from eastern Iran to the densely settled, economically strong areas of Media (where Ecbatana was a royal capital and an important mint), Elymais, and especially Babylonia. Around 108 BC Mithridates had assumed the title 'King of Kings' as used by the earlier Achaemenid rulers and the most important provinces of the empire were probably now governed by Arsacid family members bearing the title 'king'. The empire stretched from Central Asia and the Indus to the Euphrates and it was perhaps inevitable that official contact would now be made with the dominant power of the Mediterranean, Republican Rome. In around 96 BC a meeting was held between an Arsacid ambassador and Sulla, the Roman proconsul of Cilicia. What was agreed is unknown but they almost certainly discussed the contentious issue of Armenia which would remain the main area of rivalry between the two empires for the next two-and-a-half centuries.

A bloody struggle for power within the Arsacid royal family ultimately brought Orodes II (57–37 BC) to the throne and within a few years his empire faced attack by a Roman army. This was led by Marcus Licinius Crassus, who, along with Pompey and Julius Caesar, was a member of the so-called First Triumvirate that dominated Roman politics. In a division of Roman territories Crassus had been awarded Syria and in 55 BC took the opportunity to seek military glory by invading Arsacid territory beyond the Euphrates. His defeat and death at the battle of Carrhae in northern Syria with the loss of thousands of men was a serious humiliation for Rome. It also revealed the effectiveness of Arsacid cavalry – both heavily armoured (so-called cataphracts) and light brigades of mounted archers – against heavy infantry.

Yet another murderous power struggle at the Arsacid court marked the accession of Phraates IV (37–2 BC) and it forced many nobles to flee westwards. Some ended up in Antioch where Mark Antony had just arrived from Rome. In an attempt to emulate Alexander of Macedon and to take revenge for the defeat at Carrhae, Antony gathered an army of 100,000 men and marched across Armenia to Media in 36 BC. Unhappy at the slow progress he was making, Antony made the mistake of offloading wagonloads of siege engines and weapons and then proceeded to lay siege to the Parthian royal residence of Praaspa. The site has not been located with certainty but is often identified with Maragheh at the south-eastern corner of Lake Urmia. Meanwhile, Phraates led an army around the Romans and attacked and captured the baggage train, killing 10,000 men and taking prisoner several thousand more. As winter approached and a shortage of supplies started to take its toll the Romans at Praaspa found that they were unable to defend their position from attack while maintaining the siege, and so Antony was forced to retreat. Although he avoided the fate of Crassus, Antony struggled back to Syria with the loss of more than 40,000 men. He campaigned twice more in the Zagros but this came to nothing. However, Phraates was unable to consolidate these successes as his throne was threatened by an internal rebellion. To avoid an attack from the Romans while he crushed his opponents, the military standards and prisoners captured at Carrhae and Praaspa were returned to them in 20 BC. Rome, now ruled by the emperor Augustus, celebrated this as a great triumph and it ushered in a long period of peace between the empires. The Romans recognised the Euphrates as a frontier whilst the Arsacids would eventually accept a share of Roman authority in Armenia.

## PEACE AND PROSPERITY

Peace brought with it prosperity as reflected in an increased number, density, and size of settlements. This is revealed by surveys of sites in the Diyala region and parts of western Iran. In order to support the many new towns and cities, agriculture was expanded substantially with large and complex irrigation works involving canals and, in the uplands, qanats – gently sloping underground tunnels delivering water from the interior of a hill to fields below. This transformation of cultivation is especially evident in Babylonia as well as in northern, western, and south-western Iran, suggesting at least a degree of central planning. The extent of state control is difficult to determine but attempts to enforce it may have led to rebellions such as that by Seleucia-on-the-Tigris between AD 36 and AD 42. Other regions remained largely independent of the Arsacid kings. Susa (Seleucia-by-the-Eulaios) flourished under the Arsacids but Elymaean coins were struck there from AD 70 to AD 75, suggesting an expansion of the mountain kingdom which probably led to a complete unification of the highlands and Susiana around AD 120.

The wealth and sophistication of the aristocracy of Mesopotamia and Iran during the last century BC and the first century AD is probably to be recognised in the statues of stone and bronze discovered in what may have been a sanctuary at Shami – a remote valley in the Zagros east of Susa – that had been sacked in antiquity. The most remarkable piece is surely the bronze

of a man nearly two metres high and measuring a good 66 centimetres across his broad shoulders (fig.98). The figure is completely preserved except for the arms. The head, which was cast separately from the body, depicts a fairly young man with fine moustache, and full hair held in place by a wide band around the head. These features, together with the torque around his neck and his elaborate clothes, suggest that he is an Iranian (perhaps Elymaean) noble.

The economic stability that followed the ending of serious conflicts between Rome and the Arsacids was also reflected in the flourishing transcontinental caravan trade along the Silk Route. Caravans travelling east could start in Roman Antioch, cross the Euphrates at Zeugma and either follow the river south to Seleucia-on-the-Tigris or traverse the Syrian steppe to the Tigris. From the lowlands the caravans followed the traditional routes through the Zagros passes to Ecbatana before crossing the Iranian Plateau and on through Central Asia to Han China. Another important route was by boat from Babylonian ports through the Persian Gulf to India. Trade was in both directions, driven by a growing demand for luxury goods especially silks and muslins, precious stones and pearls, bronzes and glassware, oils, perfumes, and spices.

The importance of this extensive exchange network is demonstrated by the revival of the ancient centres of Nineveh and Assur which flourished again in northern Mesopotamia. The city of Labbana was built on top of the ruins of Assur and included an extensive palace with a central courtyard, dominated on all four sides by a massive, vaulted *iwan* – a large living room for working and sleeping but also where guests were received. These are distinctive features of Arsacid period palace and temple architecture, as is apparent at Hatra, another rich trading city around 50 kilometres to the west of Assur. Here the architecture and art demonstrates a much clearer fusion of eastern and western traditions: indeed it lay on an important route to the west that ran through the Syrian desert-city of Palmyra. The princes of Hatra were Arabs who were loyal vassals of the Arsacid kings. Unlike the use of traditional mudbrick at Assur, the public buildings in Hatra were constructed in fine limestone masonry on a rubble core. The city lay behind massive outer walls with a circular plan and at its centre was an impressively proportioned rectangular enclosure. The main building inside is usually identified as the Temple of the Sun where the most prominent architectural features are two large central *iwans* flanked by smaller ones, their facades decorated with engaged columns with Corinthian capitals (fig.99). The arches were decorated with sculpted friezes of human figures while carved human heads adorned the inner walls of the *iwans*. These heads gaze out of the reliefs with an emphasis on frontality rather than naturalism – an

**FIG. 98** Bronze of a noble from Shami, Iran, *c.*100 BC–AD 100, ht. 194 cm, © The Art Archive / Archaeological Museum Tehran / Gianni Dagli Orti

**FIG. 99** View of the *iwans* of the Temple of the Sun at Hatra, Iraq © By Staff Sgt. JoAnn Makinano via Wikimedia Commons

approach that is common in the visual language of the empire. However, the more naturalistic rendering of figurines known from graves may represent a continuation of earlier Seleucid traditions (see fig.93). Such composite statuettes of females are depicted nude with their left arm bent and palm outstretched (fig.100). Although the statuette illustrated here is without provenance, almost identical examples have been excavated from graves in Babylonia.

Among the most important cities in the Arsacid Empire was Ctesiphon. It may have been Vardenes I (about 40–45 BC) who had chosen this military encampment on the Tigris opposite Seleucia-on-the-Tigris for his capital, but a later Roman account credits the transformation of the settlement into a magnificent city, settled with people and surrounded by walls, to King Pacorus II (AD 78–105). The Arsacid royal court seems to have travelled seasonally along the Great Khorasan Road between the winter residence at Ctesiphon and the palace at Ecbatana. Ctesiphon soon became a target for Roman armies as tensions over Armenia led to a breakdown in the good relations that had existed between the two empires. The events are difficult to reconstruct in detail as the sources are often inadequate and always presented from the viewpoint of Roman authors – we know next to nothing of events in the east. The clear impression, however, is that the Arsacids were forced onto the defensive.

FIG. 100 Composite alabaster statuette of a female, *c.*100 BC–AD 100, ht. 20.7 cm, AN1896–1908 E.3446 © Ashmolean Museum, University of Oxford.

Armenia and frontier disputes provided the Roman emperor Trajan (AD 98–117) with an opportunity to reverse the policy of Augustus and to seek glory by emulating Alexander and invading the east. Having conquered Armenia and turned it into a Roman province in AD 114, the following year he expanded his control over the kingdoms of northern Mesopotamia. By AD 116 Trajan was ready to strike at the Arsacid capital and he led his forces south to Seleucia-on-the-Tigris and Ctesiphon which fell rapidly to him. Marching on to the shores of the Persian Gulf he received the submission of the ruler of Characene. Trajan had probably overextended his reach and Arsacid resistance forced him to withdraw, during which he laid siege unsuccessfully to the city of Hatra. Early in AD 117 he died in south-eastern Anatolia on his way to Rome. Under his successor Hadrian (AD 117–138) Roman foreign policy in the east was reversed yet again with the Euphrates re-established as the frontier between the empires. Peaceful conditions prevailed until AD 161 when Vologases IV (AD 147–191) took control of

Armenia as an Arsacid dependency following the death of the Roman emperor Antoninus Pius. A Roman counter-offensive in AD 163 won them back Armenia and two years later they moved south and captured both Seleucia-on-the-Tigris and Ctesiphon. However, the army was struck by plague and forced to withdraw in AD 166 allowing the Arsacids to reclaim the invaded territory. A pattern of Roman invasions of Mesopotamia and Arsacid counter offensives continued into the third century and must have resulted in considerable suffering for the local populations as well as damaging the wider economies. In AD 197 the Roman emperor Septimius Severus (AD 193–211) led his forces into Mesopotamia and, according to the contemporary Roman author Cassius Dio:

> *Thus he soon had seized Seleucia and Babylon, both of which had been abandoned. Later, upon capturing Ctesiphon, he permitted the soldiers to plunder the entire city, and he slew a vast number of people, besides taking as many as a hundred thousand captives. He did not, however, pursue Vologases [V, AD 191–208], nor even occupy Ctesiphon, but, just as if the sole purpose of his campaign had been to plunder this place, he was off again, owing partly to lack of acquaintance with the country and partly to the dearth of provisions. … Severus now crossed Mesopotamia and made an attempt on Hatra, which was not far off, but accomplished nothing; on the contrary, his siege engines were burned, many soldiers perished, and vast numbers were wounded.*

Severus's son Caracalla (AD 198–217) pursued the war with the Arsacids. In northern Mesopotamia he annexed the kingdom of Osrhoene and attacked Adiabene. Cassius Dio reports that Caracalla:

> *now ravaged a large section of the country around Media by making a sudden incursion, sacked many fortresses, won over Arbela, dug open the royal tombs of the Parthians [Arsacids], and scattered the bones about. This was the easier for him to accomplish inasmuch as the Parthians did not even join battle with him. … The barbarians took refuge in the mountains beyond the Tigris in order to complete their preparations …*

With Caracalla's assassination in AD 217 his successor, Macrinus, confronted the Arsacid army at Nisibis but following a bloody battle a peace treaty was agreed in which the Romans paid the Arsacid king a vast indemnity.

## THE SASANIAN EMPIRE

**FIG. 101** View of Ardashir's palace at Firuzabad, Iran ©Milad Vandaee (دالیم یئادنو) (Own work) [CC BY-SA 3.0 (http://creativecommons.org/licenses/by-sa/3.0)], via Wikimedia Commons

There may have been every expectation that this see-saw of incursions and counter attacks would have continued if it had not been for a rebellion by the Arsacids' vassal ruler of Parsa. This was Ardashir who, claiming descent from a certain Sasan, established the Sasanian dynasty. He became king of Istakhr, near Persepolis, around AD 206–208 and founded a new royal city called Ardashir Khurra, 'the glory of Ardashir' (modern Firuzabad), which expressed his imperial aspirations. In layout it formed a perfect circle with the walls pierced by four gates at equal intervals. Some distance away a large palace was constructed. It was built facing a lake and was almost certainly surrounded by a paradise garden (fig.101). At its front was a long *iwan* which led into a high-domed room, presumably where the royal throne was placed. After about ten years of rule, Ardashir began to subjugate his neighbours by military conquest. Having defeated the king of Ahwaz (possibly Elymais), he invaded Susiana, once again unifying the lowlands with the highlands. Soon afterwards his forces were in Mesene (southern Mesopotamia). By the time the Arsacid king Artabanus IV (AD 216–224) moved against him, Ardashir had the upper hand. A decisive battle was fought in April AD 224 at Hormozdgan (perhaps near Ramhormoz at the southern end of the Susiana plains) where Artabanus was killed and the Arsacid dynasty was brought to a violent end.

**FIG. 102** Rock relief of Ardashir I at Naqsh-i Rustam, Iran, showing Ardashir and Ahuramazda , *c.*AD 230–240, ht. 4.2 m © Paul Almasy/Corbis

While some of the noble families and client kings submitted to Ardashir, many others in Mesopotamia, Media, and Armenia did not and it was not until AD 226 that he was able to enter Ctesiphon and claim the Achaemenid/Arsacid title 'King of Kings'. Like Cyrus before him, Ardashir turned east and extended his control across the Iranian Plateau towards Central Asia and the Indus Valley, although the dating of these campaigns is uncertain. By the early AD 230s he was back in Mesopotamia encroaching on Roman Syria. The Romans responded by invading territory south of Lake Urmia – just as Mark Antony had done over two-and-a-half centuries before – but the outcome is unclear. The situation seems to have remained relatively quiet until around AD 238 when Sasanian forces again advanced into northern

**FIG. 103** Silver drachma of Shapur I, *c.*AD 241–270, diam. 2.6 cm, HCR7058 © Ashmolean Museum, University of Oxford

Mesopotamia and Syria: Carrhae, Nisibis, and even Hatra fell to them. To Romans like Cassius Dio it seemed that Ardashir sought to restore the Achaemenid Persian Empire:

> *He accordingly became a source of fear to us; for he was encamped with a large army so as to threaten not only Mesopotamia but also Syria, and he boasted that he would win back everything that the ancient Persians had once held, as far as the Grecian Sea, claiming that all this was his rightful inheritance from his forefathers.*

As if in confirmation of this aim of restoring the earlier Persian Empire, the Sasanian king had a relief carved into the rock face at Naqsh-i Rustam close to the Achaemenid Royal Tombs, showing his investiture by the great god Ahuramazda. Both king and god are mounted on fine horses and presented heraldically opposite one another (fig.102). Ardashir receives the diadem of kingship from Ahuramazda while Artabanus is trampled under the king's horse. In parallel, Ahuramazda's horse stands on Ahriman, the Zoroastrian embodiment of evil – thus good triumphs over evil on earth as in heaven. The figures, carved in high relief, are polished while the background is left rough to create an effective contrast. Ardashir wears not the jewelled Arsacid crown but a new and distinctive headdress consisting of a simple skull cap and with his hair gathered in a large silken globe or *korymbos*. Each Sasanian king would adopt his own personal style of crown and these help us to identify specific rulers. Inscriptions on coins issued by Ardashir describe him as a worshipper of Ahuramazda and depict a fire altar: in Zoroastrianism, fire is the son of Ahuramazda and through such imagery the king announced the union of his state and his religion.

In AD 239 Ardashir's son Shapur I (AD 239–270 or 273) had been made co-regent, becoming 'King of Kings' in AD 241, the year before his father died. The new, but experienced, king is depicted on a silver drachma wearing a crenellated crown with earflap, his hair gathered up into a silken globe (fig.103). The accompanying inscription is in Aramaic, the official administrative language of the empire: 'The Ahuramazda-worshipping

Divine Artaxerxes, King of the Kings of Iran, heaven-descended of the Gods'. On the reverse the connection with Ahuramazda is emphasised with the depiction of a fire altar and two armed attendants leaning on long staffs.

Shapur soon faced a Roman army under Gordian III (AD 225–244) that departed Antioch in spring AD 243 to reverse Sasanian advances. This was the first of three major confrontations with the Romans, from which Shapur emerged victorious and which was described by the king in a triumphant account carved at Naqsh-i Rustam:

> *Gordian Caesar raised in all of the Roman Empire a force from the Goth and German realms and marched on Babylonia against the Empire of Iran and against us. On the border of Babylonia ... a great battle occurred. Gordian Caesar was killed and the Roman force was destroyed. And the Romans made Philip Caesar. Then Philip Caesar came to us for terms, and to ransom their lives, gave us 500,000 denars, and became tributary to us ... And Caesar lied again and did wrong in Armenia. Then we attacked the Roman Empire ... In the third campaign, when we attacked Carrhae and Urhai [Edessa] and were besieging Carrhae and Edessa Valerian Caesar marched against us. He had with him a force of 70,000 ... And beyond Carrhae and Edessa we had a great battle with Valerian Caesar. We made prisoner with our own hands Valerian Caesar and the others, chiefs of that army, the praetorian prefect, senators; we made all prisoners and deported them to Parsa.*

Shapur's victories were celebrated in a number of rock reliefs. That at Naqsh-i Rustam depicts Gordian's successor, Philip the Arab (AD 244–249), in the act of kneeling before the mounted Sasanian king who grasps in his right hand the wrist of the captured Valerian (AD 253–260) (fig.104; see also fig.87). In these campaigns Shapur had captured as many as thirty-seven Roman cities including Antioch, and Valerian's son Gallienus was forced to rely on the army of the ruler of Palmyra to bring to a halt, in around AD 262, any further Sasanian invasions. Roman prisoners were deported to cities in Mesopotamia, Parsa, and Susiana, some of which – such as Bishapur on the road that linked Istakhr and Ctesiphon – were new royal foundations by Shapur. Captive sculptors and masons were put to work. Among their constructions was a 550-metre-long bridge-dam at Shushtar on the Karum River in Susiana (fig.105). Indeed, two of Shapur's rock reliefs carved in the Bishapur gorge and commemorating his victory over the three emperors may have been the work of Roman craftsmen.

Among the deportees were many Christians. There were already small Christian communities in northern Mesopotamia, Elymais, and Persis, and along with Jews and Buddhists, they were able to practise their faith freely. But this religious tolerance ended with the death of Shapur. Over a period of twenty years there followed a succession of three Sasanian rulers, all called Bahram, during which time a Zoroastrian priest called Kartir rose to prominence. He strictly enforced the religion's doctrines and led in the persecution of minority faiths. The extent of Kartir's influence over the ruling kings was made apparent by his appearance in royal rock reliefs –

**FIG. 104** Rock relief of Shapur I at Naqsh-i Rustam, Iran, showing Shapur and the defeated Roman emperors Philip and Valerian *c.*AD 260–270, ht. 7 m © Paul Almasy/Corbis

**FIG. 105** Remains of the bridge-dam at Shushtar, Iran, *c.*AD 260–270 © Livius.org

his image was even added to Shapur's relief at Naqsh-i Rustam (see fig.104, in the shadow behind the king). The instability of Sasanian leadership was also accompanied by a loss of territory to the Romans: the emperor Carus sacked Ctesiphon in AD 283 while in AD 298 the Romans under Galerius attacked northern Mesopotamia via Armenia, advancing into Adiabene before plundering Ctesiphon. A treaty imposed the following year extended Roman power to the eastern banks of the upper Tigris.

In the first decades of the fourth century Sasanian nobles ruled on behalf of the infant Shapur II (AD 309–379). By the time he had reached adulthood Shapur had gained full authority and from around AD 330 he began to re-establish Sasanian control over the eastern borders. In the west, however, he was faced by a system of fortresses in Roman Mesopotamia that had been constructed under the emperors Diocletian and Constantine and these proved to be a strong bulwark against his expansionist schemes. The rise of Christianity in the Roman Empire was also a cause for concern. Constantine had decriminalized the Christian faith in AD 312 and, according to the writings of Eusebius (Bishop of Caesarea in about AD 315–340), some eleven years later had written to Shapur urging him to protect the Christian populations in Iran. This contributed to a sense of unease about the loyalty of Christians and led to renewed persecutions: it has been suggested, though probably erroneously, that this was the cause of a rebellion by the population of Susa. The uprising was savagely crushed using elephants to flatten houses in the city.

In AD 359 Shapur attempted to push into Roman Mesopotamia but this resulted in the emperor Julian leading an impressive army against him, which reached the walls of Ctesiphon by AD 363. Julian was mortally wounded and his successor Jovian sued for peace in order to secure a safe passage for himself and his men. Remarkably, therefore, Shapur gained much of northern Mesopotamia and, eventually Armenia.

## A FLOURISHING OF THE ARTS

Although there had been repeated conflict at the borders of the Sasanian Empire and attacks on its capital, peace and prosperity had prevailed internally and this had led to a flourishing of the arts. By the later fourth century AD the use of stucco decoration had replaced royal rock reliefs, silver vessels were being produced for the court and glass ware, silks, and jewellery were being traded over remarkable distances. For example, in southern Mesopotamia at Kish two large buildings with ornamental pools and *iwans* were decorated with numerous stucco plaques similar to those also known from Ctesiphon. Many of these had geometric designs and floral motifs as well as depicting animals such as bulls, lions, and birds. The main courtyard of one of the buildings was ornamented with several stucco heads of a king (fig.106). With great bunches of curls on his shoulders and wearing a mural crown, he has been identified with Bahram V (AD 420–438), famous in Iranian legends as Bahram Gur, a mighty hunter of wild asses.

**FIG. 106** Stucco head of a Sasanian king, from Kish, Iraq, *c.*AD 420–438, ht. 47 cm, AN1932.980 © Ashmolean Museum, University of Oxford

**FIG. 107** Glass vessel with wheel-cut facets, *c.*AD 200–600, ht. 8.8 cm, AN1965.739 © Ashmolean Museum, University of Oxford

Kish also produced examples of Sasanian glass similar to the hemispherical bowl shown here (fig.107). Although without provenance, it represents a widespread type of late Sasanian glass vessel, found in excavations of Mesopotamian and Iranian sites dating to the fifth, sixth, and seventh centuries AD – examples have even been found as far afield as Japan where they had arrived as trade items. The bowl, which has lost much of its surface colour through weathering, was probably made by blowing molten glass into an open mould; the facets were then wheel-cut and polished.

Silver vessels are undoubtedly among the most spectacular works of art from the Sasanian period. Ancient authors describe how Sasanian kings gave silver plate as gifts to their nobles and contemporary rulers. From the time of Shapur II the motif of the king as hunter became a standard design and indeed, until the sixth century AD, no human figure other than the king was represented on Sasanian silver, which was probably part of an official court production. On this example, separate pieces of silver were inserted into the plate to provide high relief (fig.108). It was then gilded and inlaid with niello – an alloy of sulphur and silver – to provide the dark contrasts. The identity of the Sasanian king represented here is uncertain although the style of his crown suggests that he may be Peroz I (AD 459–484).

FIG. 108 Silver plate with an image of a Sasanian king hunting rams, *c*.AD 460–530, diam. 21.9 cm, Fletcher Fund, 1934 34.33. © 2015. The Metropolitan Museum of Art/ Art Resource/Scala, Florence

By the end of the fourth century a new threat had emerged in the east. Widespread movements of tribes in Central Asia led to their incursion into Armenia, Syria, and northern Iran. In Bactria and eastern Iran the so-called Hephthalite Huns had become the dominant power and the pressure on the Sasanian frontiers was immense. On the death of Yazdegerd II (AD 439–457) his son Peroz I only secured the throne with support from the Hephthalites. Despite this early support from the Huns much of the reign of Peroz was devoted to confronting the Hephthalites in battle in a vital tug-of-war over territory and the crucial east–west trade routes. The effect of heavy taxation to pay for these wars combined with a seven year famine must have caused terrible suffering within the Empire. To add to the misery Peroz was killed fighting in the east, along with the finest of the Sasanian cavalry, in AD 484. This was followed by a Hephthalite invasion of Iran and the imposition of an annual tribute on the Sasanians.

## KHOSROW OF 'THE IMMORTAL SOUL' AND THE COMING OF ISLAM

The former strength of the Sasanian Empire was only recovered during the long reign of Khosrow I (AD 531–579), known in Persian as *Anushiruwan*, 'the immortal soul'. Having established himself on the throne after defeating and executing his rival brothers, Khosrow concluded a peace treaty with Justinian, ruler of the Byzantine (East Roman) Empire and continued the annual payment to the Hephthalites. This provided him with an opportunity to develop further many of the internal reforms initiated by his father. Of crucial importance was a survey made of Sasanian territories and its resources, which formed the basis for a reform of the tax system that brought an assured income into the treasury. There was also a major reorganisation of the state bureaucracy and the military which established a new social class of small land owners known as *dehgans* who owed their position to the king rather than to the nobility. The empire was now divided into four military districts, each commanded by a general, with a focus on the defence of the frontiers: eastern Iran; Media and north-western Iran; Parsa, Susiana, and the Persian Gulf coast; and Mesopotamia. State sponsored irrigation works on a prolific scale helped to revive the agricultural potential of Mesopotamia: among the most impressive was the so-called Nahrawan Canal along the eastern banks of the Tigris and the lower course of the Diyala (which was established as a separate administrative district).

After his centralising reforms, Khosrow was in a position to repudiate the treaty with Byzantium and in AD 540 he invaded Syria. Antioch, the walls of which had been weakened by an earthquake about fifteen years earlier, fell to the Sasanians and the city was plundered of its riches. To house the deported population Khosrow built a new city on the outskirts of Ctesiphon which he called *Veh-az-Antiok-Khosrow* meaning 'Khosrow [built this] Better-than-Antioch.' The war with Byzantium continued on and off for around twenty years until a 'fifty-year' peace treaty between the empires was established in AD 561. This provided an opportunity for Khosrow to stop paying tribute to the Hephthalites and, in alliance with Turkic tribal groups that had moved into Central Asia, he defeated them, taking much of Afghanistan.

Trade flourished, not only along the Great Khorasan Road but also through the Persian Gulf. Sasanian control extended along the eastern shores of the Arabian Peninsula to include the wealthy states of southern Arabia, which had become Sasanian dependencies under Khosrow, and ensured a near monopoly of trade in the Indian Ocean. A hint at the resulting wealth of the empire is reflected in the remnants of the great palace that Khosrow had

FIG. 109 View of the Taq-i Kisra at Ctesiphon, Iraq, photograph taken in 1932, Creswell Archive EA.CA.6262 © Ashmolean Museum, University of Oxford

constructed at Ctesiphon. Known as the Taq-i Kisra this is but one facade of a great enclosed courtyard which had two colossal *iwans* at opposite ends (fig.109). Only one *iwan* survives with its brick barrel vault measuring 35 metres high and 50 metres long, making it the largest arch of its kind ever constructed. Roman and later Arab writers described the interior as being covered in mosaics that included an image of Khosrow at the siege of Antioch. The royal throne was set at the back of the *iwan* where the king sat in state wearing a crown of gold, pearls, rubies, and emeralds that was so heavy it had to be suspended from the top of the vault by a gold chain.

By the time of Khosrow's death in AD 579 the Sasanians had regained Mesopotamian territory east of the Euphrates and much of Armenia was back within their political orbit. Nonetheless, the struggle with Byzantium continued under his son and successor Hormuzd IV (AD 579–590), while at the other end of the empire the Turks attacked in the north-east. The latter were driven back by a Sasanian army led by Bahram Chubin, one of the leading Iranian nobles and general of Media and north-western Iran. Khosrow's military system had provided Bahram with immense authority and his victories against both the Turks and Byzantines was viewed as a threat by Hormuzd who attempted to reduce the general's influence. Bahram and other nobles reacted by rebelling against the king, who was

FIG. 110 Rock relief of Khosrow II at Taq-i Bustan, Iran, showing the king hunting wild boar, *c.*AD 591–628, ht. 4.3 m
© Paul Almasy/Corbis

murdered in a palace coup that probably involved his brothers. The forces of Hormuzd's son Khosrow were defeated outside the walls of Ctesiphon and he fled to Byzantine Syria while Bahram was crowned king. After only one year Khosrow returned with the support of Byzantine troops and defeated Bahram who fled to the Turks and was subsequently assassinated. Although now established on the throne, Khosrow II (AD 591–628) would face internal revolts for nearly a decade. For at least six years his uncle led a rebellion in the east of the empire – even minting his own coins – but he was eventually defeated. However, soon after, war was resumed with the Byzantine Empire: in AD 602 the emperor Maurice, who had assisted Khosrow II to claim his throne, was assassinated and in response, the Sasanian king launched campaigns against the new emperor Phokas, conquering much of Armenia, Cappadocia, and Syria. Khosrow also deposed Al-Nu'man III, the vassal king of the Lakhmids, an Arab Christian kingdom in southern Mesopotamia with its capital at al-Hirah on the Euphrates. The Lakhmids had maintained the defences of southern Mesopotamia against Arab tribes further west and south. With the Lakhmid buffer removed there was nothing to prevent incursions and around AD 604 Sasanian forces were defeated by an Arab alliance in the battle of Dhi Qar. Despite this, no action was taken by the Sasanians to strengthen the frontier – an omission which would ultimately prove fatal.

By AD 610 the Byzantine Empire was under severe pressure from northern invaders (Avars and Slavs) moving through the Balkans. Phokas was deposed in favour of Heraclius who made overtures of peace to Khosrow but these were rejected. Khosrow pressed his advantage and the Sasanian forces rapidly swept through the Near East: Antioch and Damascus fell in AD 612; Jerusalem in AD 614; much of Anatolia in AD 615; and finally Alexandria and with it Egypt in AD 619. It was as if the Achaemenid Empire had been re-established. The king now resurrected the tradition of royal rock reliefs. At Taq-i Bustan a monumental *iwan* was cut out of the rock face and decorated on all surfaces with astonishingly carved images of animal hunts set in a paradise park (fig.110). On the left of the *iwan* each stage of a boar hunt is shown within a single scene: the wild boar are driven to where the king waits in a boat ready to release his arrows at them.

Despite the remarkable Sasanian military advances they dissolved as quickly as they had been achieved. Heraclius successfully defended Constantinople and then went on the offensive, reclaiming Anatolia and Armenia during AD 622 and attacking north-western Iran the following year. In AD 627 the Byzantine armies pushed through north-western Iran into Mesopotamia, sacking the royal palace of Dastagird close to Ctesiphon. These victories led to a revolt at the Sasanian court and Khosrow was murdered. The nobles now installed a succession of puppet kings and even one queen, Boran, but the collapse of effective royal authority clearly reflected a weakened, war-weary empire. As such, despite being able to draw on enormous resources of manpower, the empire was unable to confront effectively the Arab armies that now began to cross the borders.

By AD 630 the Prophet Muhammad (*c.*AD 570–632) had conquered and unified much of Arabia under Muslim rule and his armies had begun to raid into southern Mesopotamia, following the well-established route along the western shores of the Persian Gulf onto the alluvial plain. In March AD 633, a year after Yazdegerd III (AD 632–651) was installed as Sasanian king, Muhammad's successor Abu Bakr (AD 632–634) sent his general Khalid ibn al-Walid with a large force into southern Mesopotamia where he achieved decisive victories against Sasanian armies in a series of battles. Khalid was then ordered to Syria giving the Sasanians an opportunity to regroup. In AD 636 a large Sasanian army confronted Muslim forces that had halted at Qadisiyya, close to the old Lakhmid capital of al-Hirah. The ensuing battle lasted five days and the Sasanians – despite probably having twice as many soldiers in the field – were defeated. The Arabs moved forward to Ctesiphon which fell in March AD 637 after a siege of about two months. Yazdegerd fled to Media and by January AD 641 Susa had fallen, leaving the Zagros as a natural barrier between the Caliphate and Sasanian territory. Continued raids by Sasanian forces into Mesopotamia prompted Caliph Umar (AD 634–644) to order an invasion of the Iranian Plateau. At the battle of Nahavand, near Ecbatana, a Sasanian army was crushed and by AD 644 territories east of the Zagros were under Muslim authority. Yazdegerd withdrew to Merv but he was rapidly losing support as the Arab armies moved eastwards. In AD 651 the last Sasanian king was killed – legend claims that this was at the hands of a humble miller but members of the local nobility were the more likely perpetrators.

Following the defeat of the Sasanian armies and the death of Yazdegerd, the Arabs faced further local rebellions but ultimately these were crushed. The new rulers of Mesopotamia and Iran continued to rely on local governors who did much to perpetuate Sasanian systems of organisation. Indeed, the *dehgans* were especially important in helping to maintain continuity at a bureaucratic level. Conversion to Islam was gradual, adopted by many for political, socio-cultural or spiritual reasons, or simply owing to persuasion. Those who converted were allowed to continue in positions of power. Christianity had flourished throughout the Sasanian Empire and its practitioners were recognised as 'people of the book' and thus tolerated and given autonomy – many Zoroastrians survived by converting to Christianity.

FIG. 111 Silver Sasanian-style drachma with Islamic inscription, *c.*AD 660–700, diam. 3.3 cm, HCR4810 © Ashmolean Museum, University of Oxford

The continuation of certain artistic techniques, such as silk production, silver plate, and stucco, can be explained by the use of Sasanian craftspeople in the service of the new state. Perhaps the most striking case of continuity, however, is in coinage: this seventh century Islamic coin is an exact reproduction of a Sasanian *drachm* but with an additional marginal phrase in Arabic *bism Allah* (In the name of God) (fig.111).

***

There are, of course, no clear-cut endings in history but this juncture has been chosen as a fitting point at which to bring the story to a close. Following the Arab invasions of Mesopotamia and Iran there was initial continuity but this should not be overemphasised, and the introduction of Islam and the Arabic language and script brought with them fundamental changes. The succeeding centuries of Muslim rule are as remarkable for their social, political, economic, and cultural achievements as anything that went before and the relationships between the populations of the highlands and lowlands played a significant part in shaping them. But these stories are told by others.

# THE QUESTION OF CHRONOLOGY AND FURTHER READING

Chronologies of the ancient Near East have been reconstructed using a combination of relative dating techniques, such as the stratigraphy of archaeological excavations, absolute chronologies, based on calendrical and astronomical records obtained from ancient texts, and scientific methods, such as radio-carbon dating and thermoluminescence. As none of these approaches are totally reliable, scholars continue to debate a range of possible chronologies, and alternative dates to those presented here will be found in the books listed below. All dates should therefore be understood as approximate, especially for the very early periods.

What follows is a small selection of reliable and accessible works that include both broad surveys and focussed publications, together with the latest interpretations and extensive bibliographies.


ARUZ, J., K. Benzel, and J.M. Evans, ed., *Beyond Babylon: Art, Trade, and Diplomacy in the Second Millennium* BC (New Haven and London: Yale University Press, 2008)

ARUZ, J., S.B. Graff, and Y. Rakic, ed., *Assyria to Iberia at the Dawn of the Classical Age* (New Haven and London: Yale University Press, 2014)

ARUZ, J., and R. Wallenfels, ed., *Art of the First Cities: The Third Millennium* BC *from the Mediterranean to the Indus* (New Haven and London: Yale University Press, 2003)

CRAWFORD, H., *Sumer and the Sumerians* (Cambridge: Cambridge University Press, 2004)

CRAWFORD, H., ed., *The Sumerian World* (Abingdon: Routledge, 2013)

CURTIS, J.E., and J.E. Reade, ed., *Art and Empire: Treasures from Assyria in the British Museum* (London: The British Museum, 1995)

CURTIS, J.E., and N. Tallis, ed., *Forgotten Empire: The World of Ancient Persia* (London, The British Museum, 2005)
CURTIS, V.S. and S. Stewart, ed., *Birth of the Persian Empire* (London: I.B. Tauris, 2005)
CURTIS, V.S. and S. Stewart, ed., *The Age of the Parthians* (London: I.B. Tauris, 2007)
CURTIS, V.S. and S. Stewart, ed., *The Sasanian Era* (London: I.B. Tauris, 2008)
CURTIS V.S. and S. Stewart, ed., *The Rise of Islam* (London: I.B. Tauris, 2009)
DARYAEE, T., *Sasanian Persia: The Rise and Fall of an Empire* (London: I.B. Tauris, 2013)
HARPER, P.O., J. Aruz, and F. Tallon, ed., *The Royal City of Susa* (New York: Metropolitan Museum of Art, 1992)
KUHRT, A., *The Ancient Near East: c.3000–330* BC (Abingdon: Routledge, 1995)
LEICK, G., ed., *The Babylonian World* (Abingdon: Routledge, 2009)
LIVERANI, M., *The Ancient Near East: History, Society and Economy* (Abingdon: Routledge, 2013)
PETRIE, C.A., ed., *Ancient Iran and Its Neighbours: Local Developments and Long-range Interactions in the 4th Millennium* BC (Oxford: Oxbow Books, 2013)
POTTS, D.T., ed., *The Oxford Handbook of Ancient Iran* (Oxford: Oxford University Press, 2013)
POTTS, D.T., *The Archaeology of Elam: Formation and Transformation of an Ancient Iranian State* (Cambridge: Cambridge University Press, 2nd Edition, 2015)
RADNER, K., *Ancient Assyria: A Very Short Introduction* (Oxford: Oxford University Press, 2015)
ROAF, M., *The Cultural Atlas of Mesopotamia and the Ancient Near East* (New York: Facts On File, 1990)
VAN DE MIEROOP, M., *A History of the Ancient Near East, ca.3000–323* BC (Chichester, West Sussex and Maldan, Massachusetts: Wiley Blackwell, 3rd edition, 2016)

# Index